Building a Thriving Future

Building a Thriving Future

Navigating the Metaverse and Multiverse

Paola Cecchi-Dimeglio

The MIT Press
Cambridge, Massachusetts
London, England

The MIT Press
Massachusetts Institute of Technology
77 Massachusetts Avenue, Cambridge, MA 02139
mitpress.mit.edu

© 2025 Paola Cecchi-Dimeglio

All rights reserved. No part of this book may be used to train artificial intelligence systems or reproduced in any form by any electronic or mechanical means (including photocopying, recording, or information storage and retrieval) without permission in writing from the publisher.

The MIT Press would like to thank the anonymous peer reviewers who provided comments on drafts of this book. The generous work of academic experts is essential for establishing the authority and quality of our publications. We acknowledge with gratitude the contributions of these otherwise uncredited readers.

This book was set in ITC Stone Serif Std and ITC Stone Sans Std by New Best-set Typesetters Ltd. Printed and bound in the United States of America.

Library of Congress Cataloging-in-Publication Data is available.

ISBN: 978-0-262-04993-1

10 9 8 7 6 5 4 3 2 1

EU Authorised Representative: Easy Access System Europe, Mustamäe tee 50, 10621 Tallinn, Estonia | Email: gpsr.requests@easproject.com

To the women who have been a compass in my life:
Catherine (my mother), Amy, Iris,
and for their supportive friendship in helping me achieve the impossible:
K. C., Hogene, Laila, Salena, Douja, Sandrine

In loving memory of Esmond, who left us too soon but forever helped bring my voice and vision to the world.

To my husband Peter, my soulmate and the love of my life, my rock.

Contents

Introduction 1

I   Journey into the Metaverse

    1   Where It Came From, What It Isn't 11
    2   The Big Prize 31
    3   The Behavioral Shifts 51

II A Blueprint for a Thriving Metaverse

    4   Unlocking Opportunity 67
    5   Charting the Path 79
    6   Designing for Impact 95
    7   Leveraging Strategic Alliances 119

III Metaversal Frontiers: Transforming Industries and Professions

    8   The Public Sector: Government, Education, and Health Care 135
    9   The Private Sector: Commerce, Banking, and Beyond 153
    10 Telecoms, Tech, and Transformation 181
    11 The Future of Work 205

Conclusion 223

Acknowledgments 227
Appendix 231
Notes 241
Index 297

Introduction

The metaverse isn't coming—it's already here. Is it the internet all over again? Yes, but it's different. And businesses that don't adapt now will be left behind. Much like the internet in the early 1990s, the metaverse is evolving at an unprecedented pace, transforming how we shop, work, and interact. The difference? This time, it's bigger, faster, and more immersive than anything we've seen before.

Where are we in this evolution? If the metaverse were following the trajectory of the internet, we'd be somewhere in the early 1990s.[1] But the key distinction today is the sheer speed of change and the scale of economic opportunity. In 2000, the global internet economy was valued at $1 trillion. By contrast, a 2022 Citigroup report projects that the metaverse economy could be worth $13 trillion by 2030.[2] This is not just hype—it signals a massive transformation in digital business, consumer engagement, and market expansion.

Yet, discussions about the metaverse tend to fall into two extremes: those hyping astronomical earnings projections and those warning about risks like digital addiction, misinformation, and exclusion. Both perspectives hold some truth. We've seen before how new technologies can reshape industries, open up new markets, and create wealth, but only when built with a clear understanding of long-term user adoption and engagement.

The metaverse is not just a technological shift—it's a transformation in how we work, connect, and create value. But a thriving future doesn't build itself. It requires strategic foresight, intentional design, and a commitment to ensuring that the next digital era is not just profitable but sustainable, adaptable, and built for broad participation. The companies that succeed won't just react to change—they will drive it, shaping an interconnected ecosystem that fosters innovation, resilience, and long-term growth. Much like the early internet, the metaverse presents massive opportunities, but only for those who recognize the need to build with purpose.

A thriving future depends on strong foundations—systems, policies, and structures that are built to last, not quick fixes that unravel under pressure. It demands collaboration across industries, disciplines, and communities to ensure emerging technologies create opportunity rather than reinforce old barriers. Companies that harness diverse perspectives and talents will be the ones that innovate more quickly, reach wider markets, and unlock new economic potential. This isn't just about ethics—it's about business strategy. When technology is designed for broader engagement, it scales faster, attracts wider adoption, and delivers higher returns.

At the same time, resilience and adaptability will separate the leaders from the laggards. The pace of technological change is accelerating, and businesses that fail to anticipate disruption—or build systems flexible enough to evolve—will struggle to compete. Sustainability will also be a defining factor. The metaverse is being built today, but its long-term viability depends on balancing economic growth with responsible development. Those who ignore these priorities risk financial and reputational fallout as consumers and markets shift toward value-driven engagement.

Ultimately, *Building a Thriving Future* is about more than just technological advancement—it's about ensuring that the digital world we create is dynamic, accessible, and built to empower. The metaverse won't wait. Companies that take an intentional, strategic approach—designing for flexibility, long-term impact, and broad market engagement—will be the ones that define the next digital era.

In the metaverse, broad market reach isn't just an advantage—it's a necessity. Businesses that design with adaptability in mind—anticipating shifts in consumer behavior and broadening accessibility—will gain a competitive edge, capture greater market share, and drive stronger revenues.[3] Talent is everywhere. Businesses that fail to attract, retain, and integrate a variety of perspectives risk stalling innovation and missing untapped opportunities. Likewise, money is everywhere. Companies that overlook key consumer segments or create rigid, inaccessible digital experiences risk alienating millions of potential users and leaving significant revenue on the table.

This isn't about corporate messaging or symbolic gestures—it's about competitive positioning, mitigating risk, and ensuring long-term market success. Consider the cautionary tales of brands that failed to understand their global audience: Pepsi's "Live for Now" commercial, Dove's "Show Us" campaign, Dolce & Gabbana's "Eating with Chopsticks" video, H&M's "Coolest Monkey in the Jungle" hoodie, and Unilever's "Fair & Lovely" rebrand all sparked widespread backlash due to miscalculations in consumer sentiment. The fallout? Damaged reputations, legal action, and revenue losses. One of the most effective ways to avoid these pitfalls is to build teams that reflect a broad spectrum of viewpoints, ensuring that critical blind spots are addressed before they become costly mistakes.

Companies that embed user-focused design into their product development, marketing, and business strategy consistently outperform competitors. Consider Nike's collaboration with adaptive athletes or Apple's commitment to seamless, intuitive design—both have driven record engagement and customer loyalty. The metaverse will follow the same trajectory. Businesses that design for flexibility, adaptability, and frictionless engagement will be best positioned to lead in this rapidly evolving space.

This book will not predict whether the metaverse will deliver its promised trillions—but will help you ask and answer the right questions:

What is the metaverse, and how does it work?

Where are the opportunities, and how do we capitalize on them?

What risks must be navigated, and how do we avoid past mistakes?

How can companies ensure they are positioned for long-term success?

How are major economic sectors applying the metaverse, and how will they continue to apply it?[4]

It will also examine the converging elements of the metaverse, the aspects that have yet to take shape, and the gaps that still remain.

This book provides the first strategic blueprint for the metaverse/multiverse(s) that empowers leaders across sectors and industries to make informed decisions and leverage the metaverse and its market expansion. *Building a Thriving Future* helps leaders form flexible, supported, day-to-day decisions and craft responsive long-range plans. It is divided into three parts:

1. Origins and trajectory: Where the metaverse came from, where it's going, and what's driving its rapid evolution.
2. Strategic framework for inclusion: How businesses can integrate scalable, user-driven design into metaverse platforms, workforce strategies, and technological innovation.

3. Industry impact: A deep dive into how different sectors—government, health care, finance, and technology—are adapting and why inclusion is key to their success.

Even as the metaverse takes shape, some still ask if it is real or viable. Meanwhile, what was once science fiction has now become virtual reality (VR), augmented reality (AR), and mixed reality (MR). Emerging technologies depend on other elements within a technology ecosystem—such as more users with headsets or the broader rollout of high-speed networks. Chatbots like Bard, ChatGPT, and Nova AI did not spring up overnight. Even when technology is viable, economic feasibility and consumer readiness dictate success. And consumers are ready.

One of the key reasons the metaverse is more viable today than ever before is the rapid advancement of artificial intelligence (AI). Without AI, the metaverse would remain a fragmented, impractical concept. AI is the invisible engine behind real-time user interactions, natural language processing, automated content creation, and adaptive digital environments. Generative AI tools now enable users to create immersive worlds, digital assets, and lifelike avatars without extensive coding expertise, democratizing content creation in ways previously unimaginable. Machine learning powers recommendation systems, virtual assistants, and predictive analytics, shaping personalized, intuitive metaverse experiences. Meanwhile, AI-driven speech recognition, real-time translation, and emotion analysis are removing barriers to participation, making digital spaces more accessible to users across languages, cognitive styles, and physical abilities. Just as search engines and social media required AI to become seamless parts of daily life, the metaverse's evolution depends on AI's continued refinement. Businesses that understand this intersection—and act on it now—will be the ones shaping the next era of digital transformation.

The metaverse has been priming consumers for years. Video games have introduced players to avatars and immersive multiplayer experiences since 1974. Over the decades, technological advances—from 3D movies to interactive media and AR/VR games—have made virtual experiences richer. Platforms like *Minecraft* and *Second Life* built virtual worlds with thriving communities long before the metaverse became mainstream. *Pokémon GO* familiarized millions with augmented reality. Today, consumers are not just gaming in the metaverse—they are shopping, learning, and engaging in commerce. Consumers have grown accustomed to the convenience of digital interactions—whether for shopping, education, banking, or health care. The 3D immersive metaverse promises to re-personalize and rehumanize these processes, restoring the social dimension that traditional online interactions lack.[5]

Economic forecasts predicting the metaverse's multitrillion-dollar potential assume it will be broadly accessible—designed to work across global markets and technological infrastructures. Failing to meet these expectations could stall adoption, spark consumer backlash, and limit financial returns. Businesses that wait to refine their approach until after missteps occur risk lost opportunities and reputational damage. The lesson from past digital revolutions is clear: the companies that anticipate shifts in consumer behavior, design for broad market adoption, and create seamless engagement will be the ones to lead.

The shift toward digital-first economies has already changed how businesses approach their markets. Decades of consumer insights have shown that understanding different audiences and tailoring experiences to their expectations drive engagement and revenue growth. Thoughtful, strategic design in product development and digital experiences not only strengthens brand reputation but also prevents costly miscalculations. Misreading consumer sentiment has resulted in high-profile failures—campaigns that were intended to be bold but instead sparked backlash, leading to PR crises and financial setbacks. Businesses must take a proactive approach, ensuring that their digital strategies align with market expectations, cultural awareness, and technological adoption patterns.

Despite the complexity of building the metaverse, user-centric design and market-driven accessibility are not just possible—they are essential. This book will highlight the economic advantages of expanding digital experiences to wider audiences and provide strategic guidance for integrating these principles across innovation, product development, and industry applications.

Failing to prioritize broad usability in a hyper-connected world limits market potential and revenue growth. The speed with which a competitor can enter the metaverse with a more seamless, widely accessible, and user-driven approach should serve as a wake-up call for business leaders. Many companies are already designing digital ecosystems with scalability, adaptability, and cross-market appeal in mind. Those that do not risk falling behind.

Much of the insight in this book is shaped by my experience, which spans over two decades in academia at both Stanford and Harvard, my consulting work with major companies in technology and nontechnology sectors, and my advisory role with international organizations like the United Nations and other governmental bodies worldwide focused on the use of technologies, AI, and big data in the labor market.

As a researcher, strategist, and practitioner at the intersection of AI, big data, workforce development, and organizational transformation, my career has been dedicated to equipping leaders with the insights they need to make better, data-driven decisions. Over the past two decades, my work has focused on helping organizations navigate complex change by leveraging AI, big data, and behavioral science to refine strategies, drive innovation, and optimize talent management.

Through my company, People Culture Data Consulting Group, my teams and I have worked with businesses to integrate AI and big data into decision-making processes, demonstrating how these technologies can transform systems and improve outcomes. Without these tools, companies operate in the dark—reacting instead of anticipating, adjusting instead of leading. AI, big data, and data analytics provide the ability to recognize patterns, identify emerging challenges, and implement solutions rooted in evidence rather than assumptions.

My expertise is further shaped by my role at Harvard University, where I apply AI and big data to study digital transformation and the evolution of work. AI is no longer confined to academic research or specialized industries—it has gone mainstream. The rapid adoption of generative AI across sectors is accelerating the integration of the metaverse into daily life. While some still see the metaverse as a distant concept, its impact is already tangible, reshaping industries and redefining engagement.

In my consulting work, I have seen companies struggle to harness the full potential of their workforce, often due to communication breakdowns or unintended barriers to collaboration. Even subtle missteps—like the language used to encourage innovation—can have unintended consequences. For example, I've seen organizations use terms like "harvesting" to describe the collection of employee ideas, not realizing that such phrasing can discourage participation. A single misplaced word can disrupt an entire innovation pipeline. As businesses shape their metaverse strategies, ensuring clarity in communication and engagement will be critical to fostering an environment where creativity and problem-solving thrive.

The behavioral science approach to leadership, growth, and innovation that I present in this book has been tested and refined in real-world business environments. The strategies and insights provided will help leaders confidently navigate this evolving landscape, equipping them with practical tools to adapt and implement these principles effectively.

Additionally, my work with the United Nations' International Telecommunication Union (ITU) has given me a global perspective on the evolving digital economy. As vice chair of the steering committee for the ITU Global Initiative on Virtual Worlds and AI: Discovering the Citiverse, and co-chair of the ITU's Committee on Accessibility and Inclusion, I have contributed to shaping the international dialogue on metaverse governance. My involvement in the ITU's Focus Group on the Metaverse (FG-MV), which was established in December 2022 to create a standardized roadmap for metaverse adoption, has reinforced the importance of global collaboration in designing digital ecosystems that are scalable, sustainable, and aligned with future economic and technological realities.

The ways in which the metaverse can be monetized are limited only by creative innovation. Revenue streams will emerge from both digital and real-world products and services. Established brands will harness AR and VR to market and distribute physical products, while consumers will engage with fully digital assets and immersive experiences. The potential spans everything from virtual concerts and digital real estate to next-generation telemedicine and interactive retail, opening up unprecedented business opportunities.

Success in the metaverse requires a well-executed strategy that balances technological innovation with user engagement. Businesses—aware that they are building interconnected digital ecosystems—must begin to see their users as active participants rather than passive consumers. Beyond securing brand loyalty and revenue growth, companies must ensure that their virtual environments are seamless, engaging, and built for long-term adoption. A metaverse designed with adaptability, accessibility, and consumer trust at its core is not just a competitive advantage—it is a requirement for sustainable success.

The metaverse has already moved beyond science fiction and into multiple realities. It may seem new today, but it won't for long. If the metaverse expands, scales, and integrates into mainstream digital life as projected, even dominant industry players may find themselves struggling to compete with agile, early adopters. Businesses that fail to act now risk being outpaced by those that recognize the urgency of innovation and the importance of designing for long-term market viability. Armed with the right strategies and insights, leaders can confidently position their organizations for success in virtual, augmented, and mixed-reality environments.

In the time it has taken to read this introduction, something new and significant has already happened in the metaverse economy. A company has launched a breakthrough product, a startup has redefined an industry, and a new insight has pushed the boundaries of what is possible. The metaverse is not waiting for businesses to catch up—it is evolving in real time. The companies that anticipate shifts, embrace digital-first strategies, and create user-driven experiences will define the future of this space. This book will be your guide to navigating these changes, making informed decisions, and unlocking opportunities in this new digital frontier.

I Journey into the Metaverse

1   Where It Came From, What It Isn't

The word "metaverse" conjures a mix of images from virtual reality to online gaming to new social media tools. In fact, the metaverse consists of a series of technological convergences that are already blending digital and physical worlds. It's easy to assume, mistake, and mythologize. Nonetheless, we are able to identify what the metaverse is, what it is not, how it got here (and it is here), and where it may be headed in the near future (and you can bet it's moving). Understanding where the metaverse comes from can inform business decisions and help organizations find their competitive positions.

COVID-19 disrupted traditional work and learning environments, affecting workers in offices, cubicles, call centers, and classrooms. Many employments retreated to the safety of remote work locations, and the metaverse welcomed a significant boost. During the COVID-19 lockdowns, immersive virtual activities gained popularity. Improved and affordable VR (virtual reality) headsets and AR (augmented reality) glasses, along with the availability of virtual experiences, created new possibilities. E-commerce experienced significant growth, and when Facebook changed its name to Meta, curiosity and interest exploded.[1] But, the metaverse was up and running long before then. Just ask gamers and Second Lifers. From the launch of *Second Life* in 2003 to Microsoft Mesh in 2021, metaverse usership has steadily grown.[2] Each day, the metaverse marches closer to being another new normal. For more than two decades, *Second Life* has allowed users to create customized avatars and interact in various environments. The platform is used for socializing, entertainment, education, business, and collaboration, with registered accounts reported at just over 73 million, and daily active users between 800 and 900 thousand.[3] Users have embedded themselves in the *Second Life* metaverse for more than two decades, and that is only one such platform—others already exist and many more are coming.

A Dream and a Process

So, where did our perception of the metaverse begin? As stated, the word "metaverse" evokes images of avatars—digital representations of users—participating in activities from shopping to attending concerts to gaming. There is, however, a backstory. Humans have been blending worlds of fantasy and fiction with real experience for a long time. In the early nineteenth century, stereoscopes, holography, and the View-Master (introduced in 1939) made efforts to create 3D and rudimentary immersive experiences.

Looking at any evolutionary process, it can be difficult to believe that one thing really led to another. Just look at mobile phones, personal computers, or processes like filing taxes or making purchases. The primitive internet of the early '90s and its early search engines and web browsers like Archie and Mosaic are the ancestors of the metaverse. Advanced as these tools were when they were first introduced, it took time to catch up with tech predictions of video telephony and flip phones seen in *The Jetsons* and *Star Trek*. The 1987 *Star Trek* reboot popularized the holodeck, "where images of reality . . . created by [the ship's] computer [were] highly useful in crew training [and] highly enjoyable when used for games and recreation."[4] Our metaverse, used for these same purposes, is still working to achieve holodeck quality.

What's in a Name?

"Metaverse" is a blend of the words "meta," referring to a transcending nature (i.e., "beyond"), and "verse," a shortening of "universe," illustrating a domain that extends beyond our physical reality.[5] The metaverse is typically perceived as a multidimensional virtual space where physical and nonphysical elements integrate. At its core, it involves a digital amalgamation of numerous virtual environments. Within these crafted digital realms, people partake in a spectrum of activities, from work to recreation, either singularly or in groups. It is foreseen that individuals might lead discrete, parallel existences across various universes within the metaverse, transitioning beyond their physical selves into alternate digital realities. The metaverse transcends the status of merely a virtual space, becoming a crucial, transformative entity in individuals' daily experiences.

Straight Outta Fiction . . .

The intricate timeline of the metaverse spans over three decades, crucially intertwining with diverse technological advancements and popular culture. The term "metaverse" was coined in 1992 in Neal Stephenson's novel, *Snow Crash*.[6] Its conceptual framework dates back further, evolving alongside digital twins, blockchain technology, and decentralized autonomous organizations, introduced in 2002, 2009, and 2016, respectively. *Neuromancer* by William Gibson, a 1984 novel in the same genre, popularized "cyberspace" and "cyberusers."[7] Prior to the use of the term "metaverse," the 1978 introduction of MUD1, the first multiplayer real-time virtual world, and the 1982 release of *Tron*, the first movie to depict a digital world, offered glimpses of what a metaverse could be like.

The *Matrix* movies depict human existence within a virtual reality. *Second Life* has allowed users to live in a virtual world for two decades. The Roblox multiplayer game platform, introduced in 2006, now has over 55 million users.[8] Platforms like *Second Life* and Roblox began to materialize various forms of the metaverse, offering users unprecedented abilities to create avatars and navigate through virtual worlds. The increasing sophistication of these platforms, as witnessed by Roblox's growth for example, suggests a trajectory toward a future wherein the metaverse is deeply integrated into our digital interactions.

Popular culture and literature have galvanized interest in the metaverse. Ernest Cline's *Ready Player One*, published in 2011 and adapted into a film by Steven Spielberg in 2018, introduced a wide audience to a conceivable digital future, foregrounding the concept of the metaverse. Across these novels and films, however, underlying dystopian elements are evident. These cautionary tales warn about the impact of political and social trends, offering insight into the ominous feelings that accompany the myth that the metaverse and artificial intelligence (AI) will consume and diminish humanity.

How It Started

Beginning with MUD1 in 1978 and Ultima Online in 1995, gaming has played a significant role in jump-starting the metaverse.[9] Games like *World of Warcraft* have provided immersive virtual experiences for decades, helping to popularize the concept of virtual worlds where people can socialize and interact. Gaming has also driven the development of VR headsets like the Oculus Rift and HTC Vive, making 3D VR more feasible and affordable. Gamers may have been some of the first consumers to see the possibilities of the metaverse, dedicating significant time and effort to its acceleration.

At the same time, gaming has sparked concerns about the extent of human activity in the metaverse. Leaders question if the metaverse is technologically ready to function. Some people see the metaverse and its attendant AI as sinister, wondering if they will replace real people and real life.[10]

Replacing Real Life?

They won't. This is one of many myths about the metaverse that we can debunk: The metaverse will not supplant real-life human interaction. In fact, the metaverse is an analog of the real world. It relies on the real world for its shape and content. No real world, no metaverse. Eventually, the metaverse will exert a shaping influence on the world that created it. But the metaverse is still very new. All the ways in which it can be used have not been determined.

As the metaverse is rolled out piece by piece, it will offer new approaches, options, and strategies. Businesses in sectors impacted by the technology will have to pivot or face potential loss. Various companies such as Walmart, Gap, Hulu, and Adidas, have already entered the metaverse, establishing their virtual presence.[11] Business leaders have gone from asking *"What is that?"* to *"What do we do now?"*

Myth vs. Reality

Before we answer that, let's list a few more myths, some of which we will debunk throughout this book. When the word "metaverse" popped out of science fiction into technological fact, it brought along quite a few myths and misconceptions. Many mistakenly believe that the metaverse, with its cohort AI, will replace the real world, as discussed above. Others incorrectly assume that the metaverse is simply another word for VR or that it is one unified space. Another myth holds that large tech companies will monopolize the metaverse. The fear that it will be a dangerous, dystopian realm also contradicts the emerging reality.[12]

Using the expansion of the internet from its inception to today as a rough gauge, companies and users are expecting to spend considerable amounts of time in the metaverse. Presently, consumers may spend an average of 0.3 hours per day in the metaverse. But by 2030 they are projected to spend at least 2 to 4 hours in the metaverse daily, with an average of 2.5 hours minimum per day.[13] This shift signals a major economic opportunity for those who understand the metaverse's potential and move early to build sustainable, adaptable business models within it.

More Than Just Novel Tech

So, what *isn't* the metaverse? It is not just some advanced technology. It is more than the tools and equipment that enable immersion.[14] When people think of the metaverse, the primary image that arises may be that of someone wearing AR/VR equipment. The metaverse also conjures up gaming, which formed the basis for much of the technological development. But it is more than play. While the emergence of Web 3.0—the next phase of the internet, characterized by decentralized networks, blockchain technology, and greater user control—offers the technological infrastructure and flexibility to power the metaverse, it is not simply Web 3.0. The metaverse offers a new and authentic digital approach to all our activities.[15]

With its origins and name rooted in science fiction, metaverse activities and experiences already feel unlimited. Networks and infrastructure, supporting hardware, and services are rising to meet the vision of the metaverse. While the metaverse of fiction evoked a dystopian vibe, there is no plan to roll out an unhealthy metaverse. The immersive 3D virtual and augmented worlds can be designed to positively impact user and community well-being.

Technology intent on helping people create and move about in an alternate or augmented reality has taken several forms, arguably over centuries. Now that it has become part of the possible, consumers and experts seem to be scrambling to say what it is and where it came from.[16] The popular, amusing genre of "What People Think I Do" memes fit the metaverse with striking precision. What do people think? Some people think the metaverse will trigger conflict and damage society, some think it will replace their reality, some think it will be dystopian and addictive, some think it will make them millions of dollars in a very short time, some think it will make humans obsolete, and many know that it's just an extension of existing reality that provides opportunities for more lifelike, personalized, immersive experiences.[17] Consumer clarity will help the metaverse move through its origin stories and land in households and workplaces around the world.

For the Few, the Young, the Tech-Savvy

People have many questions about the metaverse: what it is, who will use it, how long it will last, and how it will be monetized. Many think it's just for gamers or that only young people will use it. These are myths.

Its userbase is already multigenerational and diverse. Pioneering users and early adopters expect to be in the metaverse for years to come. They expect to use it for increasing numbers of activities across their daily routines. The ways in which individuals and businesses will derive revenue from the metaverse will be limited only by the breadth of organizational and individual imagination.

Remember that consumers anticipate that within five years they may be spending at least 2 to 4 hours in the metaverse daily.

Other metaverse myths focus on usage and users, assuming it requires specialized equipment that only certain people have or are accustomed to using in gaming culture.[18] And while headsets are preferred for a fully immersive experience, they are not necessary for everything in the metaverse. In reality, you can access it using devices like computers, smart glasses, and smartphones. In addition, this myth overlooks the many users who plan to engage the metaverse not for gaming but for re-personalized, immersive education, shopping, social interaction, travel, and more. The metaverse may be a natural next step for Gen Z, but millennials report a greater awareness of the existence of the metaverse than other generations.[19] Furthermore, data indicates that multiple generations are poised to engage the metaverse, and several are already donning headsets and exploring.

But the Tech Hasn't Hit Yet

Assumptions about the metaverse and its limitations often focus on technology and economics. VR headsets and AR glasses are available, but they are not yet mainstream. Perhaps by the time you are reading this book, Meta and Ray-Ban have already launched a new version of their original product (available since 2023) or the new Google Vision Pro headset has been released.[20] It is therefore a myth that the technology required to bring the metaverse to market is still years in the making. This error can incline leaders to put off their plans and projects. This myth seeks to explain why things are moving as slowly as they are. Business leaders who embrace this assertion and structure decisions on this basis risk being left out of formative experiences and initial windfalls. Economic viability is another concern. The metaverse holds significant promise for e-commerce, offering new tools like responsive, customized avatars for sales and customer service, and introducing digital assets like virtual real estate and NFTs. These assets are already a significant part of metaverse economies. As the metaverse grows, opportunities for economic gain will expand, requiring leaders to adapt and innovate continuously.[21]

The metaverse can create new experiences, connect people globally, and enhance existing activities, presenting opportunities and threats to traditional business models. Tech companies are particularly well positioned, providing the platforms, applications, and content that drive the metaverse. Essential elements like AI, 5G networks, and blockchain enable seamless, immersive experiences. Innovations in operating systems, devices, and accessories will shape individual experiences, supported by a secure, high-speed infrastructure.[22] Basic processes such as encryption, authentication, monitoring, biometrics, cryptocurrencies, digital wallets, and payment processors will also play crucial roles, offering further business opportunities.[23] Despite myths, the metaverse is already attracting significant investment, doubling in the past two years.

As the metaverse evolves, AR and MR (mixed reality) tools provide immediate opportunities for users to explore and acclimate, blending real-life environments with digital content and paving the way for broader adoption.

Right into Business

The path to the metaverse culminated in a major industry shift in 2021, when Facebook rebranded as Meta. This move brought the once speculative concept into the mainstream, sparking widespread interest and highlighting the industry's consensus on its potential.[24] Microsoft's launch of Mesh and other tech giants' ambitious plans further emphasized the momentum toward metaverse development.

Governmental and institutional entities have also embraced the prospective future within the metaverse. From the city of Seoul releasing its five-year metaverse plan to the UN and International Telecommunication Union (ITU) initiating research and pilot projects in 2023, the metaverse has begun percolating through various layers of society, industry, and governance, moving beyond its initial conceptual domain into a future that, while still being molded, is steadily hurtling toward reality.[25]

In line with these efforts, the ITU's Global Initiative on Virtual Worlds and AI: Discovering the Citiverse (for which I sit as a vice chair of the steering committee) has taken a leading role in shaping the governance and implementation of metaverse solutions in urban contexts.[26] The Citiverse Initiative serves as a global platform that encourages the creation of open, interoperable, and innovative virtual worlds for people, businesses, and public services to engage with safely and effectively. It also facilitates the development of normative frameworks, focusing on principles, enablers, and governance, that will guide the integration of metaverse solutions within cities. By providing training, organizing events, and offering a sandbox environment for testing metaverse scenarios, the initiative is helping cities worldwide explore the potential of the metaverse, contributing to the global effort of shaping its future. During the 2024 World Telecommunication Standardization Assembly (WTSA-24) in New Delhi, India, ITU members approved Resolution 105, marking a significant milestone in promoting and strengthening metaverse standardization—a first within the UN system and having global implications.

As our cyber and physical realities become more intrinsically entwined, it becomes evident that the metaverse is not transient. It is a burgeoning realm, gradually being fortified by technological advancements and becoming entwined with our digital interactions, offering new possibilities for socialization, business, and governance in a unified digital space. Consequently, strategic discussions are perceptibly shifting from considerations of the metaverse's longevity to meticulous planning regarding short- and long-term strategies as we navigate through this digital metamorphosis.

Progress and Phases

The unfolding journey of the metaverse spans from its initial emergence, to an advanced stage between 2024 and 2027, and foreseeably culminating into a mature phase from 2028 onward.[27] AI has already been pivotal in shaping the metaverse, but the advent of generative AI marks a substantial advancement.[28] This new iteration accelerates the process of turning ideas into incredibly realistic content and experiences, representing a critical evolution for businesses and investors alike. A pivotal underpinning, AI ensures the robustness of metaverse infrastructure, offering crucial data for the higher echelons and playing an indispensable role in creating virtual environments for social interactions, as evidenced by technologies from Nvidia.

Amid its evolution, external factors such as the war in Ukraine or Israel–Palestine conflict have influenced the market, affecting supply chains and altering demand dynamics, particularly in eastern Europe, illustrating that the development of the metaverse does not occur in isolation but rather is interwoven with global events and markets.

Metaverse? Multiverse(s)? Metaverses?

The terminology around the metaverse is evolving. Fifth-century BCE Greek philosopher Democritus proposed an infinite number of universes, each with unique laws and properties.[29] Today, each major metaverse platform, such as Decentraland, The Sandbox, Meta Horizon Worlds, VRChat, NeosVR, Sansar, Somnium Space, and Star Atlas, operates as a distinct metaverse. The term "metaverse" encompasses both the concept and these discrete virtual realms, supported by various platforms. This idea aligns with the concept of the multiverse, where multiple interacting metaverse platforms coexist.

While often perceived as a shared space, the metaverse is more accurately a network of interconnected virtual platforms. Each business building a virtual platform contributes to the creation of numerous, often interlinked metaverses, catering to sectors like Industry 4.0 and entertainment. Industry 4.0, often referred to as the Fourth Industrial Revolution, encompasses the integration of digital technologies such as the Internet of Things (IoT), AI, robotics, and data analytics into manufacturing and industrial processes. These metaverses are expected to function as united spaces, emphasizing the importance of interoperability and enabling seamless interaction across platforms and industries.

Like the internet of Web 2.0, the metaverse is built on networks, servers, infrastructure, and content—a blend of hardware and software. It offers a wide range of activities, services, games, and workspaces, many experienced in 2D. However, interactions in the metaverse, such as chatting with an assistant or making a purchase, will feel more immersive and live than in Web 2.0. Visiting landmarks or natural wonders with a VR headset offers an engagement level previously inaccessible.

The metaverse is designed to complement or supplement reality, not replace it. Although it may seem distant, the rapid adoption of generative AI by the public and organizations has accelerated its integration into our daily lives. Despite being built by various entities, the metaverse will offer seamless interoperability, allowing users to navigate different virtual environments effortlessly. Unified by underlying technology, the metaverse will function much like the laws of physics unify the universe, enabling safe exploration of otherwise inaccessible environments and offering many new features.

Enter the Avatars

The metaverse offers deeply engaging, multisensory experiences, allowing users to explore virtual realms through avatars. These avatars—digital representations of users—facilitate interactions within these artificial environments.[30] They serve as digital proxies, enabling users to express and immerse themselves in new and limitless ways. Avatars can range from idealistic replicas to entirely distinct entities, providing a canvas for innovative self-expression.[31]

For example, a user might choose an avatar that reflects their unique style or fantasy aspirations, such as a character with radiant, ethereal skin or an outfit that transcends current fashion norms. This avatar, capable of nuanced gestures and facial expressions, acts as a conduit for the user's emotions and interactions in the digital cosmos. Conversely, businesses might use avatars aligned with their brand ethos. A high-tech firm might adopt a sleek, futuristic humanoid figure, while a nature-focused enterprise could use an avatar that blends with the natural environment, adorned in earthy tones and exuding serenity. These corporate avatars, beyond visual representation, are designed for customer engagement, product demonstrations, and virtual assistance, bridging the brand and its clientele in the virtual space.

Avatar identities in the metaverse reflect personal evolution, changes in preferences, and significant life events. They provide a safe space for self-exploration, allowing users to explore different aspects of their identities and gain deeper self-understanding.[32] By bypassing societal prejudices related to physical, sensory, or cognitive traits such as appearance, age, race, gender, and disability, avatars can promote diversity and inclusion. These digital identities are the primary medium for interaction, communication, and social endeavors within the metaverse, fostering emotional safety and comfort in virtual environments.

Avatars cater to temporal self-representation, allowing users to reveal various facets of themselves at different times, perhaps embodying past, present, or future selves. They encourage an experimental and playful atmosphere, offering users the opportunity to explore aspects of one's identity, such as gender, age, race, disability, and species in a virtual context. However, it is crucial to recognize that exploring identities related to disability, race, or culture can raise significant ethical issues, such as cultural appropriation or insensitivity. Avatars are vital in preserving user privacy and anonymity, protecting against unwanted disclosure of personal data. Ultimately, avatar identities unlock limitless avenues for creativity and innovation while providing contextual adaptability, allowing users to modify their avatars for different scenarios, whether a formal professional meeting or a casual social event.

Retrofit a Competitive Edge

As a virtual world, *Second Life* illustrates both the risks and opportunities of digital platform design. In its early years, the platform catered primarily to tech-savvy early adopters, but its lack of broader accessibility and user-friendly controls limited engagement and slowed mainstream adoption.[33] Many users found the navigation unintuitive, while others faced challenges ranging from technical accessibility barriers (e.g., it was difficult for people with visual impairments or limited mobility to use) to hostile or unwelcoming social dynamics.[34] These factors restricted the platform's ability to expand into a broader market and sustain long-term growth. Recognizing these shortcomings, Linden Lab, the company that created *Second Life*, gradually retrofitted its platform to be more user-friendly, accessible, and commercially viable.[35] The company implemented simplified controls, enhanced content moderation, and built in tools for business collaboration and e-commerce. It also introduced market-driven accessibility enhancements to expand its user base. These adjustments helped stabilize the platform but came at a high cost in both time and revenue—issues that could have been avoided had these considerations been part of the initial design. For today's metaverse innovators, the lesson is clear: retrofitting accessibility, engagement, and usability after launch is costly and inefficient.

Bake in Strategic Design

Unlike the retroactive approach that *Second Life* had to take, new platforms like Microsoft Mesh have the opportunity to bake in inclusion from the outset. Microsoft Mesh, which uses Azure holographic services, is an immersive MR platform that supports connection and collaboration. Users can access the experience through head-mounted displays (HMDs), laptops, and smartphones to socialize, conduct meetings, collaborate on product designs, provide technical support, and engage in learning environments.[36] As Mesh is still under development, major issues have not yet surfaced, but the platform's inclusive design is already being prioritized to prevent potential problems.

Microsoft, aware of the potential risks of hostile or discriminatory virtual environments, has proactively developed guidelines for the creation and use of Mesh environments. These guidelines are intended to promote inclusion and respect for all users, ensuring that no group is alienated. Moreover, Microsoft is working on tools that enable users to report harassment and abuse directly within Mesh, demonstrating a forward-thinking approach that contrasts with *Second Life*'s retroactive solutions. By embedding inclusivity and accessibility at the core of its design, Mesh is positioning itself to create a safer, more welcoming metaverse for all participants.

Expanding Market Leadership

Metaverse teams shaping hardware, software, and virtual experiences have a unique opportunity to define future digital economies. Companies that prioritize adaptive and inclusive design, intuitive user interfaces, and AI-powered personalization will be able to capture new audiences while enhancing long-term retention. For example, research shows that 41 percent of women have used a primary metaverse platform for over a year compared to 34 percent of men. Additionally, female leaders are 20 percent more likely than their male counterparts to spearhead multiple metaverse initiatives. Yet, a staggering 90 percent of leadership roles within organizations defining metaverse standards are held by men.[37] This disconnect between user engagement and decision-making power presents a missed business opportunity. Companies that align leadership structures with actual and potential user demographics will have a competitive advantage in shaping digital strategies that resonate with a broader consumer base.

From Passive Users to Active Stakeholders

The companies that will lead the metaverse revolution aren't just designing virtual worlds—they are crafting dynamic accessible and inclusive digital economies. The platforms that remove barriers to participation, integrate scalable technology, and facilitate seamless collaboration will be the ones that attract businesses, developers, and consumers alike. Inclusion is not just about representation—it's about designing platforms where users become active participants, cocreators, and decision-makers. Companies that embed accessibility and inclusivity into user experiences, content creation tools, and governance structures will drive deeper engagement and innovation. A metaverse that reflects diverse perspectives will naturally generate broader adoption, greater user investment, and stronger network effects. A thriving metaverse isn't just a replication of real-world structures—it's a chance to rethink how digital spaces drive engagement, innovation, and profitability. The leaders who recognize this shift early will be the ones who define the future of digital interaction.

Learn the Language

"Web browser," "social media," "search engine," "e-commerce," "influencer," and "hashtag" were not always part of our lexicon. "Chat" and "troll" have taken on whole new meanings. But we've learned them all. Nothing signals the arrival of a new frontier like new language. Words have the power to include, exclude, inform, confuse, explain, and simplify. They often guard the entrance to fields of study, academic disciplines, and professions. In the same way, a growing lexicon of baffling terminology seems to guard the gateway to understanding the metaverse. For leaders, making decisions requires that they understand the things and functionalities behind this collection of potentially intimidating metaverse lingo such as "decentralized," "blockchain," "digital wallet," and "tokens" (fungible and non-fungible).

Once you can separate your real life (RL) from VR, AR, and MR, you have got a handle on the various modes of extended reality (XR or xR). Metaverse lingo will proliferate as new names emerge for new devices, services, and ways of doing things. Metaverse language proficiency will help leaders understand the technology, communicate with others, monitor the competition, and recognize or create opportunities.[38]

Off the Blockchain

The word "decentralized" frequently arises in discussions about the metaverse, symbolizing interconnected user-generated content and the interface enabling this interconnection and interaction.[39] These conceptual gaps between diverse digital worlds are practically bridged using blockchain technology, the digital evolution of the traditional block ledger. Essentially, blockchain is a distributed ledger where transactions are recorded in blocks securely and transparently, and this ledger is spread across a network, enhancing its tamperproof nature.

For example, when a user in the metaverse creates a unique digital artwork, the details of this artwork are recorded on the blockchain. This record includes the date of creation, ownership, and the artwork's unique attributes, ensuring its authenticity and origin are permanently logged. Such a system ensures that when the creator decides to sell this artwork to another user, the transfer of ownership is securely and transparently recorded on the blockchain, making it easy to verify and track.

While blockchain might be clouded by its association with volatile financial tools and practices, its role in the metaverse is pivotal.[40] It extends to the economic structure, influencing how value is created and exchanged. Businesses and innovators in the metaverse, rather than creating fixed, controlled content, are cultivating a decentralized space that mimics the dynamic and free environments of the real world. Blockchain enables and manages this decentralization, allowing the multiple worlds within the metaverse, each with its own digital landscape, to become interconnected.

Consider Sarah, a hypothetical user who owns a collection of digital assets in the metaverse—a unique outfit for her avatar and a virtual artwork. These items, represented as tokens in her digital wallet, can be transported across different worlds within the metaverse thanks to the interoperability enabled by blockchain. This system ensures seamless movement and recognition of assets across various digital environments, a feature that would be impossible in a fragmented, centralized system.

Blockchain not only makes these worlds inter-travelable but also adds an element of security and commercial rights management.[41] For instance, if a world within Sarah's metaverse ceases to exist, her property, securely stored in her digital wallet, remains safe. She can then transport this property to other worlds and even sell it, a capability still evolving but made possible through blockchain technology.

The decentralization powered by blockchain in the metaverse mirrors the dynamic, choice-rich environments of our physical world. It facilitates a unified yet diverse digital universe where creation, ownership, and transfer of value are secure, transparent, and centered around the user's experience.

Token Gestures and Smart Contracts

Digital tokens, enabled by blockchain, secure ownership powers for users in the metaverse. Tokens represent data units on a digital ledger. Ethereum, one of the most widely used blockchain platforms, supports both fungible and non-fungible tokens (NFTs). It is the most trusted platform for minting and trading NFTs—unique digital assets that are central to the metaverse—due to its security, scalability, and large developer community.

Tokens represent items that are assigned a fixed value or cash equivalent.[42] Real-world tokens and their digital or cryptographic cousins (like Bitcoin) are fungible. Fungible tokens on the blockchain resemble currency and can be used for purchases and payments. Like currency, the value of a fungible token (such as Ethereum, Tether, or USDC) can fluctuate with market dynamics. A fungible token worth $75 can buy a $75 item or settle a $75 debt, and a $1000 token is equal to any combination of tokens worth $1000.

In contrast, NFTs are non-fungible.[43] Each cryptographic token is unique and irreplaceable. NFTs represent unique objects in the metaverse, certifying ownership of artworks or custom-crafted avatar accessories like jewelry. If an item is certified one-of-a-kind by an NFT, only the uniquely identified object can be returned or transferred. Authenticity of fungible and non-fungible tokens is secured via smart contracts, stored on the blockchain and written in code, allowing automatic execution when conditions are met.

Smart contracts can create decentralized applications (dApps) that require no central authority.

Ethereum Request for Comments 20 (ERC-20) is the standard for fungible items, covering exchangeable in-world currencies in the metaverse.[44] ERC-721 smart contracts apply to unique digital items, covering NFTs. These programmable contracts certify ownership and uniqueness, enabling sales and the transfer of certain unique rights.

This technology transfers real-world activities to the metaverse. Decentralization, blockchain, fungible, and non-fungible tokens will influence how the metaverse mirrors the real world. NFTs will shape how companies and brands build and sustain relationships with consumers.[45] Notably, metaverse interactions may impact real-life relationships and vice versa. For example, a consumer engaging with a brand's virtual store in the metaverse might receive personalized recommendations or exclusive offers redeemable in physical stores. This symbiotic relationship enriches the consumer experience both virtually and physically, blurring the lines between metaverse interactions and real-life benefits.

Realities

And speaking of RL, the metaverse seems to offer several nuanced realities.[46] There is AR and there is VR, and these sit under the umbrella of XR. This broad term covers all varieties of computer-generated or modified reality. Under XR, VR signifies an experience where the user is fully immersed in the digitally generated environment. Content in this setting may be built from real-world phenomena using video and other media, or it may be entirely computer-generated. Often, a combination of content is used to create MR, or mixed reality. Currently, whether for work or play, accessing the full VR experience requires the use of headsets and haptic controllers.

We turn instead to AR. Augmented reality blends purely digital content with existing environments. Think *Pokémon GO* and many other applications designed for smartphones. AR starts and remains grounded in the real world and overlays digital content. MR then brings that blend of real-world and digital content to the headset. Users interact, in real time, with digital images and content but perceive them to be in their own environment. This mode can be useful for planning or creating 3D experiences of ideas or designs. It's also popular in exercise and fitness applications.

Terminology in Action

To recap, if you are in the metaverse and know yourself as an avatar, you are ready to operate in VR, AR, or XR.[47] Virtual worlds supported by Web 3.0 and protected by blockchain enable the use of cryptocurrency, a form of digital or virtual currency secured by cryptography. Unlike traditional currencies issued by governments, cryptocurrencies operate on decentralized networks based on blockchain technology, allowing for secure and transparent transactions. Keep the crypto in your metaverse wallet (Alpha Wallet, Coinbase Wallet, or MetaMask) where it can be used to purchase digital assets, participate in decentralized experiences, or trade within compatible metaverse marketplaces. Buy, sell, or exchange NFTs on Nifty Gateway, Open Sea, or Rarible. Turn your digital assets into tokens, including original creations using services like Sensorium Galaxy, and take your place in the creator economy. Welcome to the internet of value (IOV). Here, minting is the process of making new NFTs. Decentralization aims to remove centralized authority and spread trust across larger groups of users. Try play-to-earn games, gamble in virtual Vegas, or turn your virtual assets into real dollars. They are, after all, pretty phygital, just like digital twins.

If all of this gets a little overwhelming, you can retreat to a walled garden. Similar to a closed platform, a walled garden offers a closed digital ecosystem governed by separate rules and provides visitors with additional privacy and security. If "player one" is primed and ready, lingo learned, it may be time for a MMORPG (massively multiplayer online role-playing game) like *Fortnite* or *Minecraft*, where upward of millions of people can interact and play in shared spaces.

Of all the metaverse lingo, arguably the most important piece of jargon is not even a whole word. The most important metaverse term is a prefix: *inter-*, as in interoperable, integrated, and interactive. The metaverse is not quite itself without that decentralized, borderless quality. Significant technology and programming enable the characteristic seamlessness that is the ultimate goal of businesses and users alike.[48] At the beginning of inter- is "in-," as in inclusive. That defines the metaverse that is being designed and invented every day.

Metaverse lingo will proliferate as new names emerge for new devices, services, and ways of doing things. Becoming fluent in metaverse language is part of leading and decision-making in metaverse businesses. Leaders should know that the metaverse is not myth, hype, or just another digital gaming space; multiple generations of consumers are aware and ready to be all-in.

The Adoption Timeframe

The metaverse heralds a revolutionary phase in user engagement within virtual environments, evolving to offer advanced interactions, vivid imagery, realistic avatars, and an enveloping auditory experience. Projections anticipate that by 2026, a quarter of individuals will dedicate at least one hour per day to varied activities in the metaverse, spanning education, health care, arts and design, infrastructure strategizing, and manufacturing.[49]

Although the internet struggled in its infancy, there are factors that suggest that the metaverse may be adopted more quickly. The technology gap and feasibility issues are less daunting. We complain about latency, but we have high-speed networks. There is enough tech—smartphones, PCs, headsets—as well as infrastructure, platforms, and content to get going and start experimenting with the metaverse. Major tech companies such as Meta, Google, and Microsoft are building and contributing to the metaverse. Non-tech businesses including Nike, Adidas, and Wells Fargo are giving consumers a reason to jump into AR and VR. The metaverse offers novel modes of personal experiences that many consumers had to relinquish when they embraced the convenience of the internet.

Consumers across generations are developing their own notions of metaverse myth and reality.[50] These ideas may vary across generations, from Gen Z and millennials who are born wired, to Gen X and baby boomers who got the upgrade and are up to speed. Cultivating engagement from each group will require thoughtful, customized strategies. Given the goals of businesses to gain and sustain usership, busting the myths represents an essential step in removing the barriers to success.

We Have to Wait

Some experts point at the earliest to 2030 and at the latest to 2040 as the pivotal point in the growth of the metaverse.[51] The 2030–2040 predictors include venture capitalists, social media and gaming CEOs, early metaverse platforms, and other business leaders. They expect several of the enabling technologies, userbases, markets, and economic structures to be in place or mature. By 2040, it is expected that the investments being made in the development of metaverse tools and tech will have become highly profitable. Fields such as learning, health care, mental health counseling, community services, entertainment, and travel are expected to draw people into the metaverse, creating rich and dynamic experiences.

However, some argue that the metaverse may not become essential to daily life, with users still preferring real-world experiences. Concerns include potential misuse for control, exploitation, misinformation, addiction, violence, and virtual trafficking. Infrastructure development and hardware acquisition challenges could slow metaverse expansion.

Despite these concerns, generations are ready to immerse themselves in the metaverse, expecting it to become mainstream. This contradicts the myth that no one is familiar with the metaverse; people know what it is and plan to participate significantly. For businesses, this means they cannot wait. Consumers are already testing hardware, content, and services, and companies must do the same.

Businesses designing and building the metaverse have meaningful insights. Game developers, researchers, philosophers, and tech experts understand its potential. However, consumers seem to have the best grasp of its future, eagerly anticipating shopping, connecting, learning, and accessing health care in the metaverse. Their activities guide where leaders should focus to create success. Leaders must recognize that the metaverse is not just hype or a digital gaming space; multiple generations are aware and ready to engage fully.

2 The Big Prize

First, why is everyone talking about the metaverse, and where is all this heading? The metaverse is gearing up to be an expander of possibilities in all areas of human activity, so it should come as no surprise that it is also preparing to be a massive creator of value. McKinsey & Company estimates that the metaverse could be worth $5 trillion in economic value by 2030 (a number we will return to).[1] Goldman Sachs previously estimated a potential $8 trillion market by 2025.[2] PricewaterhouseCoopers (PwC) forecasted $1.5 trillion by 2030.[3] Bloomberg Intelligence projected $800 billion by 2024.[4] And Gartner anticipated $280 billion by 2025.[5] Although some of these projections will be tested or revised as we move through 2025, they remain useful markers of the scale and momentum surrounding the metaverse. Meanwhile, Meta continues to invest more than $10 billion annually in advancing the metaverse ecosystem.[6] This rapid growth and investment has propelled the metaverse into public consciousness, generating both excitement and skepticism. In 2022, investments surged, fueled by major moves such as Microsoft's $69 billion acquisition of Activision Blizzard and smaller but significant contributions, including $12 to $14 billion in venture capital and private equity. This high-profile acquisition is part of a larger trend where key players in the digital and virtual spaces are consolidating, which, in turn, is intensifying interest in the metaverse.[7]

All of this leads to the question that faces all businesses. After it was launched in 1991, one concern that hung over the internet for several years was: How do I get money out of this? Once that was answered, the internet popularized terms like "e-commerce" and "monetize." While "e-commerce" can be traced back to instances as early as the mid-1980s, it gained widespread recognition in the mid-1990s as online businesses began to emerge and flourish. Right now, leaders want to know how the metaverse will translate into stable, regular, significant revenue streams for their businesses.

Businesses are waking up to the myriad ways in which the metaverse will pay. Not surprisingly, there are more possibilities and pursuits than NFTs and crypto. A focus on value potential and creation can help businesses weigh decisions and choose exciting paths into the metaverse.

As old and new products and services vitalize the metaverse, how do businesses capture their share of the trillions of dollars that are forecast to flow? Inclusive design stands out as a key element and factor. When businesses mindfully create interfaces, content, approaches, and service models that are attractive and welcoming to everyone, they naturally position themselves to attract and sustain diverse customers.

A Few Projections

Let's start with a few projections. With the metaverse recognized as a reality, the next question is its potential value.[8] By 2030, significant investments should confirm the expected economic growth. In 2021, investments in the metaverse reached $13 billion, soaring to over $120 billion in 2022.[9] The unprecedented expectations and opportunities are clear. Consumers, accustomed to remote work and increased online shopping during the COVID-19 pandemic, are excited about the metaverse. Many companies are positioning themselves as early adopters. While estimates vary—ranging, as we saw in the chapter introduction, from $280 billion by 2025 to $8 trillion by 2025, and $1.5 trillion by 2030—McKinsey's $5 trillion projection by 2030 offers a middle ground that balances the more conservative forecasts with the more optimistic ones. This projection reflects a broader consensus on the potential for the metaverse to shape future economic landscapes, though the path to reaching that figure will depend on how investments in technology, infrastructure, and virtual platforms evolve in the coming years, especially as artificial intelligence (AI) has received more attention in recent years.

What It Does for People

The metaverse expands possibilities, reducing people's sense of limitation. With minimal equipment, users can explore new worlds or enhance their current environments. This has a significant psychological impact, akin to the early days of social media but on a faster trajectory. By the time you read this book, the metaverse may be an $800 billion-a-year market.[10]

Web 3.0, which underpins the metaverse, enhances sensory and immersive experiences beyond previous technologies. The metaverse's evolution won't be linear; breakthroughs will occur as businesses collaborate and compete to offer superior user experiences. Platforms like Roblox exemplify how user-generated content drives engagement and revenue, allowing players to create and monetize their own games. Meta's acquisition of Oculus, a leading virtual reality (VR) technology company, positions it as a major VR platform, highlighting the growing importance of immersive experiences.

Sectors like fintech (financial technology), NFTs, cryptocurrency, fashion, entertainment, and consumer goods are increasingly integrating with the metaverse, contributing to its complexity and appeal. Fintech enables secure transactions, while NFTs allow for ownership of unique digital assets. Fashion brands launch virtual clothing lines, and entertainment companies host virtual events, illustrating the metaverse's capacity to generate value and revolutionize digital interaction.[11]

Designed and Built By?

Building a thriving metaverse starts with strategic investment in emerging technologies. Yet despite massive financial investments and ongoing technological advancements, the metaverse still faces critical challenges in accessibility, representation, and market inclusivity. Over the past five years, a small segment of companies has captured the majority of metaverse-related funding. Men-led companies have secured 90 percent of total investments—$107 billion—while women-led companies have received only 10 percent, totaling $5 billion.[12] This funding gap is an economic one. It impacts product design, user experiences, and business opportunities, often reinforcing a limited perspective rather than fostering a digital space designed for global engagement.

Without a broader and diverse spectrum of investors, creators, and decision-makers, companies risk innovating in an echo chamber, leading to products that cater to narrow demographics rather than a diverse user base. For example, avatars designed without input from a diverse user base may fail to represent the broad range of identities, preferences, and needs that define digital interactions today. Currently, 60 percent of girls and 62 percent of women report feeling misrepresented in digital spaces.[13] With women making up 42 percent of global internet users, there is a substantial opportunity for businesses that prioritize adaptive, user-centric design.[14]

The metaverse also presents structural access challenges that could hinder broad adoption. Individuals with limited digital literacy, disabilities, or unreliable connectivity face barriers to full participation. AR (augmented reality) and VR technologies require significant bandwidth, and in many parts of the world, broadband access remains inconsistent. Without scalable solutions that improve accessibility, the metaverse could deepen existing digital divides rather than bridge them.

Creators of the Ecosystem

Building a thriving metaverse also starts with broadening the pool of its creators. While men have historically dominated the technology sector, shaping the early metaverse, women now make up 24 percent of its developers—an increase, but still far from parity.[15]

AI serves as a driving force behind metaverse innovation.[16] Machine learning (ML), a key subset of AI, allows systems to adapt and improve over time, while large language models (LLMs) enhance user interactions through more natural and intuitive communication. The metaverse, envisioned as a persistent 3D digital world, is powered by these technologies, from AI-driven physics simulations to adaptive virtual assistants that enhance user engagement.

Yet, without intentional and ethical design, these systems risk reinforcing or even amplifying existing biases.[17] A metaverse that reflects unconscious bias—whether in AI-driven avatars, content moderation, or algorithmic recommendations—could exclude entire groups of users or misrepresent them. Ensuring these technologies are neutral, adaptable, and reflective of a diverse global audience is not just a social goal but a business imperative for long-term adoption and growth.

A thriving metaverse requires more than just gender diversity—it demands the participation of creators from a wide range of backgrounds. Developers with disabilities, neurodiverse perspectives, and varied racial and ethnic identities bring invaluable insights into how digital environments should function. A metaverse designed by a cross-section of global talent is one that will attract, retain, and engage a truly global audience.[18]

Ecosystems

Technologies and innovations no longer stand alone; they exist in ecosystems where their success depends on multiple factors.[19] The metaverse, with its interdependencies, compels businesses to periodically revise their strategies. VR headsets' quality, convenience, and cost depend on materials and hardware, while blockchain technology shapes transactions. Emerging regulations will reshape metaverse activities, requiring businesses to examine all possibilities to choose the right path.

A diverse technology ecosystem leads to more inclusive products and services. Diverse teams of developers and designers consider a broader range of user needs and bring different perspectives, fostering creativity and innovation. This inclusivity results in products that are welcoming to all users. Innovations like screen readers, text-to-speech, closed captioning, multilingual services, and culturally sensitive products have made technology more accessible and inclusive.

Notable examples of diversity-driven innovations include emojis, which have become mainstream for effective emotional communication in digital messages. Voice assistants like Siri, Alexa, and Google Assistant have made technology accessible to people with disabilities. Social media platforms like Facebook, Twitter, Instagram, TikTok, and Pinterest have provided marginalized groups with avenues to connect and share experiences.

In the metaverse, taking an ecosystem approach helps organizations avoid developing isolated innovations. Understanding what other players are planning and capable of introducing helps create sound strategies. Narrowing the field and selecting inventions for commercialization requires assessing the ecosystem to determine product viability.

Latency (the amount of time it takes for data to get from one place to another) and rendering (the process of generating realistic images from data) impact the quality of the metaverse user experience. High latency can be disorienting, disrupting immersion, while efficient rendering is crucial for realistic environments. Content providers remain confident that improvements in Web 3.0 networks will enhance these aspects.

Research shows substantial gaming engagement among global internet users aged 16 to 24, with 91.2 percent of females and 92.4 percent of males actively playing on various devices. Overall, nearly 83 percent of internet users worldwide engage in gaming, indicating a broad and inclusive gaming culture.[20]

In the nascent stages of the metaverse, perspectives vary: 50 percent of respondents understand it and envision their organization's participation, 26 percent grasp the concept but are uncertain about involvement, and 24 percent understand it broadly but lack detailed knowledge.

Where's the Money?

Let's look at what consumers indicate they are looking forward to doing in the metaverse.[21] Users expect to socialize, shop, travel, access entertainment, and game individually or in groups. In addition to equipment and costs of connections, these activities can be monetized and represent significant revenue sources that are expected to grow over time. The immersive 3D environment promises to improve the quality of each of these activities. While it might echo earlier sections, this repetition underscores the multifaceted potential of the metaverse in enhancing user engagement and business opportunities.

From product development to training, businesses also have practical visions of the ways in which they will leverage the metaverse. These use cases are tied to earnings and expenditures. Some uses of the metaverse can potentially reduce operating costs, while others represent significant increased revenue opportunities. Meetings in the metaverse add a beneficial dimension to the 2D videoconference. Many companies believe that business processes such as recruitment, interviewing, onboarding, continuing education, professional development, promotion, and mentoring can be done better in the metaverse. For example, imagine a virtual job interview scenario where both the interviewer and the candidate, represented by avatars, interact in a more engaging 3D setting than a traditional video call. This enhanced interaction could lead to better assessment and decision-making processes. Research and development (R&D), product design, project management, collaboration, and innovation all transition well to the metaverse, especially for individuals separated by significant distances.

During COVID-19 lockdowns, events and conferences fled the in-person setting for synchronous videoconferencing and asynchronous venues. These modalities dispensed with the sensory value of face-to-face encounters but added measurable physical and psychological safety. In addition, they reduced or eliminated the need to travel. The metaverse can return some of the lost sensory components through its immersive environment and effectively re-personalize routine online activities. Because the metaverse provides seamless mechanisms for marketing and sales, businesses can leverage it to create value or derive revenue. For instance, a fashion brand could set up a virtual store where consumers can try on clothes with their avatars, offering an immersive shopping experience far beyond what current internet platforms provide.

Platforms offering customized avatars, like Facebook, Snapchat, and Microsoft have been preparing users for the metaverse. Many users now desire more interactive content, and the metaverse can fulfill these expectations. Digital tools like Bitcoin and Ethereum will support its development, while increased network hardware processing speeds will reduce latency and improve the quality of the immersive experience.

Gaming and recreational platforms are investing in the metaverse to grow usership and enhance their 3D media quality. Reliable data on consumer expectations and readiness guide business leaders to focus their efforts, creating value and turning opportunities into tangible outcomes.

Approaches to capturing value in the metaverse exist, but we can expect these to change or be reinformed as initial and later testers of the metaverse apply and revise their methods. As in the RL (real-life) world of business, having a clear goal at the outset must be a first step. Well-articulated goals and a clear sense of the role a business or individual will assume in the creation of value underlies effective strategy at any scale.

Testing, Testing

Following the goal-setting and planning, piloting the product or service allows for testing, feedback, and adjustment. This phase can resemble other approaches in digital environments, but the metaverse may introduce other variables or present different lessons. Monitoring and analyzing results may be a perennial element of the process as organizations adopt and adjust. Businesses can expect to keep refining products and processes beyond the test phase as part of an ongoing competitive strategy.

Typically, when the test phase confirms that a service or product is ready for wider rollout, the business has already lined up the essential resources and begun to integrate the efforts into the culture. The technological capabilities are where they need to be. Essential talent is in place, and the piece of the metaverse represented by the product or service has become a seamless part of the company's strategy and day-to-day operations. It's time to scale. Since the metaverse becomes the new, superseding computing paradigm, it must be ready and working right out of the box.[22] Expectations are high. Approaches to goal formation, planning, piloting, and a large-scale release must be well executed in order to secure broad adoption.

Nothing indicates the magnitude of those expectations more than the 7,200 percent jump in Google searches for "metaverse" or the 923 percent increase in investment in 2022.[23] The metaverse predicts not only the emergence of massive new economic opportunities but also the creation of novel jobs yet to be imagined. Simultaneously, the best and most challenging thing is that the metaverse and its economic potential are still in the formative stage. The overwhelming majority of businesses expect to prosper in the metaverse economy. More than half of consumers who are aware of the metaverse are already using it. These factors and many others support the prediction of $5 trillion by 2030.[24] But, the metaverse is still in its infancy. It is complex and evolving. Competition is unclear and there are still technical challenges. So, the money expected to flow from the metaverse can seem indefinite and elusive.

Here's the Money

Existing products and services offered in the immersive metaverse setting are poised to be major winners. The metaverse is expected to create significant value across mature economic sectors like health care, financial services, retail, consumer goods, technology, and manufacturing. By 2030, it is anticipated to influence 80 percent of commerce and host half of all live events virtually.[25]

Nearly 80 percent of metaverse consumers have already made purchases, and some leaders project that 15 percent of their earnings will come from the metaverse within the next five years.[26] To tap into this prosperity, anticipation and preparation are crucial.

One way companies can realize value in the metaverse is through digital twinning, which involves replicating physical objects, systems, or beings in the virtual environment. For example, a car manufacturer like Tesla could create a dynamic, interactive digital twin of a car model to run

The Big Prize

simulations and tests, experimenting with new designs or modifications without the need for physical prototypes. For instance, it could test the impact of different materials on the car's aerodynamics and fuel efficiency in a variety of virtual environments and conditions, which would be costly and time-consuming to replicate in the real world. This method allows for rapid iteration and experimentation, significantly reducing development time and costs.

Digital twinning changes how organizations operate by increasing flexibility and expanding the scope of what can be achieved. It offers unprecedented opportunities for learning and hands-on training, providing product designers, innovators, engineers, and leaders with a significant edge.

Simulations, modeling, testing, refining, and problem-solving become much more efficient in the metaverse. For instance, architects and urban planners can create detailed virtual models of buildings and cities, simulating various scenarios and making real-time adjustments. In health care, medical professionals can engage in highly realistic simulations of surgical procedures, accelerating learning and proficiency. Tech companies can test and refine new apps or software within the metaverse, allowing rapid prototyping and efficient problem resolution.

Combining talents and facilitating collaboration in the enterprise metaverse enhances organizational connectivity and decision-making. Although this ideal is still evolving, digital twinning is already a standard element in many businesses.

All about Metanomics (Meta Economics)

The digital economy is evolving into the virtual world, leading to the concept of "metanomics." Metanomics studies economic principles and activities within the metaverse, encompassing virtual and real-world goods and services.[27] It focuses on how traditional economic principles like supply and demand or scarcity apply in a virtual setting where dynamics can differ significantly.

Metanomics is essential to grasp the predicted economic value of the metaverse, which includes a mix of purely virtual products, services, and events, as well as real-world events brought into the virtual realm. This new economic landscape features virtual offerings such as fashion items for avatars, digital real estate, virtual concerts and plays, education, workforce training, and health care. A substantial portion of value will also come from marketing and selling real-world products in the metaverse, from everyday consumer goods to big-ticket items like vehicles and real estate. The metaverse can remove barriers and fundamentally alter interactions and experiences.

The novel economic dynamics within the metaverse require us to understand and apply a modified set of economic principles. Metanomics provides a fresh approach to analyzing business outcomes and economic behavior in the metaverse, considering its intersection with and differences from traditional economic systems. This approach is necessary as the metaverse introduces new methods and opportunities for economic activities, applying to wholly digital properties and products.

Metanomics is still evolving as technology finds expression in both expected and unexpected ways. The metaverse replicates aspects of the real world, and a comprehensive metaeconomy must be inclusive and accessible to all. Therefore, understanding metanomics is crucial for businesses, consumers, investors, and others as they begin transacting in AR and VR, navigating this new digital economy.

Metaverse Demographics

By 2030, the global user base of the metaverse is projected to reach between 1.4 billion (conservative estimate) and 2.6 billion (liberal estimate), with the US hosting 197 million users, reflecting a 35.5 percent penetration rate.[28]

Worldwide, user penetration is expected to rise from 14.6 percent in 2024 to 39.7 percent by 2030.[29] In the US, the penetration rate is anticipated to increase from 20.5 percent in 2024 to 56.0 percent by 2030. In 2030, the UK, Canada, Brazil, Germany, and India are expected to see user bases of 15.37 million (22.2% penetration), 16.76 million (42.6%), 66.66 million (30.4%), 19.37 million (23.2%), and 270.50 million (18.7%), respectively.[30] These figures highlight significant disparities in metaverse adoption, influenced by a combination of technological and demographic factors.[31]

Despite this rapid growth, 62 percent of American adults remain unclear about the metaverse's purpose, underscoring the need for greater education and awareness. A March 2022 survey of US adults revealed differing perceptions of the metaverse. While 12 percent expressed strong interest and 24 percent showed some interest, racial disparities emerged: 52 percent of Hispanic and Black adults indicated interest, compared to 46 percent of white adults who expressed little or no interest.[32]

In terms of awareness about the metaverse, a notable gender disparity also exists. While 17 percent of men reported having heard a lot about the metaverse, only 5 percent of women expressed the same level of familiarity. Additionally, the lack of awareness is more pronounced among women, with 50 percent reporting they had heard nothing at all about the VR environment, compared to 28 percent of men.[33]

This highlights a significant gap in metaverse knowledge and exposure between genders, which could exacerbate an existing divide that should be narrowing, not widening.

Clothing, Housing, and Other Essentials

The metaverse replicates RL, adds new elements, and extends possibilities, creating opportunities across various market areas beyond gaming. Consumers are expected to identify and acquire metaverse essentials, from practical tools enhancing virtual experiences to luxury items signifying status within the metaverse.

If programmers are painstakingly diligent and mimic reality, an avatar's clothing and shoes should gradually wear out or suffer damage during activities, creating a need for new attire.

Users may want to change outfits or acquire event-appropriate apparel. Virtual department stores and specialty clothing shops already have a presence in the metaverse, mirroring real-life shopping experiences. Spending money on virtual clothes may seem challenging to understand, but it's similar to valuing fashion in RL—about expression, identity, and status. Custom, unique items, verified by NFTs, convey exclusivity and status.

Virtual real estate has become significant in the metaverse, with platforms like Decentraland, an Ethereum-based platform, allowing buyers to purchase and develop land. Property values rise and fall, and real estate investors are actively participating.[34] Options include buying and holding, developing, or leasing/renting properties. Financing and mortgages in the metaverse are set to embrace options not utilized in the real world, accepting tokenized assets such as artwork or clothing as collateral.

In addition, a large component of metanomics will be driven by work in the metaverse.[35] Virtual events, like concerts, require expert lighting and sound, creating opportunities for meta-roadies and other virtual gig workers. Earning possibilities extend beyond obvious real-world equivalents, inviting new opportunities across borders. For example, a tech startup in Silicon Valley could collaborate with a metaverse design studio in Tokyo, reaching a global audience without physical constraints.

Gaming, virtual sports, betting, and gambling are expected to grow exponentially, with cryptocurrency introducing new options for casinos and sports betting worldwide.

Business-to-business relationships and transactions will thrive in the metaverse.[36] Companies can create or outsource VR replicas of their products or processes, enhancing R&D and innovation. For instance, a car manufacturer might work with a metaverse company for virtual test drives, offering a cost-effective approach to product development.

The metaverse promises to reduce economic gaps by providing training and job opportunities for workers from developing countries, allowing them to work for companies worldwide.

Overall, the metaverse will re-personalize many currently type-and-click-only activities, including internet gambling and betting. The reintroduction of sensory input may impact user behavior in unpredictable ways.

The Metaverse Economy

Some of the metaverse economy will be driven by activities within the metaverse itself; some of it will be driven by efforts to make the metaverse better and better. Streamlining processes, improving connections, and reducing the required equipment are innovation imperatives that enrich the metaverse economy. Building more lifelike avatars and environments will draw in businesses and individuals. Building, delivering, and refining will be as critical as selling and learning in the metaverse.

Governance and privacy have significant roles to play in the metaverse economy. Confirming identities, protecting transactions, securing assets, determining what laws apply and responding to violations are only some of what might be expected to matter. Partnerships between business and traditional governance and enforcement agencies will provide the framework for policies that enable companies to thrive and sustain necessary background security for users across the metaverse.

Operating on blockchain technology, the metaverse includes both real and fantastical realms, introducing new economic dynamics. NFTs provide blockchain-based proof of ownership and authenticity for unique items like digital clothing and artwork.[37] This parallels real-world certification but is tailored for the virtual world. Robust security and verification systems are necessary to prevent forgery, a challenge in both the real and virtual economies.

Virtual real estate in the metaverse can be bought and sold like physical property, influencing market dynamics. As the metaverse personalizes banking and financial services, it reshapes our understanding and application of metanomics, helping navigate new ventures and guiding business dynamics and innovations. Leaders should recognize the novel yet familiar economic dynamics of the metaverse, encompassing transactions within the metaverse and extending from real life into AR/VR.

Integrating the metaverse with future technologies will drive demand for higher-quality graphics and competitive consumer pricing, spurring competition and innovation. This evolving economic landscape presents regulatory challenges integral to the metaverse's growth.

The metaverse economy is ready to expand, and businesses are adapting. Web 3.0 offers possibilities beyond the metaverse, attracting investor interest in AR- and VR-based economics. As the digital world adopts these gradual changes, excitement tempered with caution characterizes the mood.

The implications for multiple sectors are substantial. Metanomics will encompass both what is sold and what is saved by engaging the metaverse.[38] Businesses will open new channels for marketing and selling goods, innovate new products and services, and realize significant savings by leveraging AR and VR for onboarding, training, and upskilling employees. As technologies mature, innovation reaches an optimal pace, users adopt, and businesses expand possibilities, metanomics as a discipline will evolve.

Balancing the Economy

A balanced economy is essential, particularly in the rapidly evolving metaverse. The emergence of high-value digital assets, such as the prestigious CryptoPunks NFT collection, sheds light on broader economic dynamics shaping virtual markets.[39] While the metaverse presents new financial opportunities, it also risks reinforcing familiar market disparities if left unchecked.

A closer look at NFT trading trends since their mainstream explosion in August 2021 reveals pricing inconsistencies linked to perceived identity traits. For example, female and darker-skinned CryptoPunks frequently command lower market prices compared to their male and lighter-skinned counterparts. Data from past sales highlights this gap: one study found that the median minimum sale price was 95 ETH for female avatars versus 99 ETH for male avatars, with 1,165 female sales compared to 2,124 male sales. Similarly, mid- and dark-skinned CryptoPunks have historically underperformed in weekly average sale prices relative to their lighter-skinned equivalents.[40]

While the influence of gender bias on NFT valuations remains variable, racial bias has proven to be a statistically significant factor. Lighter-skinned avatars often achieve higher prices, reinforcing the notion that even in digital marketplaces, visual identity affects valuation. As the metaverse expands, such imbalances, if left unaddressed, could perpetuate existing economic disparities in new digital spaces.

Scarcity and perceived uniqueness also drive value in the NFT economy. This is exemplified by the $12 million sale of one of just nine "alien" CryptoPunks at Sotheby's, demonstrating how exclusivity can significantly inflate asset prices. Yet, the same market mechanisms that reward uniqueness can also reinforce biases—whether consciously or not—by mirroring traditional value judgments from the physical world.[41]

To foster a metaverse economy that is both innovative and fair, a proactive and multilayered strategy is essential. Transparent algorithmic valuation must be implemented to ensure that digital assets are priced objectively, preventing factors like gender or skin tone from influencing their market value. Industry-wide diversity benchmarks, backed by regular audits, can help promote inclusive digital assets and balanced representation. Decentralized governance structures should empower a broad range of stakeholders, ensuring that decision-making is not dominated by a single group. Economic incentives can further encourage creators to develop diverse collections, broadening participation in the NFT economy. Regulatory and policy frameworks, developed in collaboration with metaverse platforms and policymakers, can establish fair-trade regulations that prevent discriminatory pricing patterns. Raising awareness through education and community initiatives can shift norms, fostering a marketplace that values fairness and responsible valuation practices. By integrating these strategies into the growing framework of metanomics, the metaverse has the potential to set new standards for transparency, balance, and financial inclusivity, ensuring long-term growth, stability, and innovation.

The Initial Phase

A business's initial steps in the metaverse involve learning and adjusting, providing space to understand the impact on users and communities. Early engagement, even if cautious, offers multiple advantages.

Automobiles didn't replace horses overnight; disruption took time, with some carriage makers even transitioning to the new industry. Similarly, as the metaverse approaches, business leaders face questions about why and how to get involved, what to keep, let go, and focus on, and how to plan and reap benefits.

The existing internet is to the metaverse what horses and buggies were to early cars. The internet won't disappear immediately; it will have users for a long time. Businesses and consumers engaging with the metaverse now will test new possibilities and shape success as others follow. Early adopters will define good design, set policies, build momentum, and guide the transformation.

Learning from History

The metaverse is not just a culmination of improvements to one tool or invention but the result of many efforts. AR and VR stem from experiences we've long desired. With its trail of loosely connected predecessors, the metaverse represents our best effort so far. Considering the history of the internet, the metaverse can likely impact numerous markets faster than Web 1.0 and 2.0 did.[42]

Leaders can leverage the hindsight of the internet's emergence to avoid repeating mistakes. The innovations that led to the World Wide Web took decades. From the first computer connection at MIT in 1965, using a low-speed dial-up telephone line—considered the first (albeit small) wide-area computer network—there have been numerous milestones leading up to the development of HTML and the public introduction of the web in

1990–1991. While virtual reality may seem new, it actually has a long history in fiction and science.

In 1838, Sir Charles Wheatstone's mirror stereoscope used mirrors to create a sense of depth and immersion, marking the birth of VR according to some experts. In literature, Stanley Weinbaum's 1935 short story "Pygmalion's Spectacles" featured goggles that inserted the wearer into a movie with active senses. In 1962, Morton Heilig's Sensorama booth combined several technologies for an immersive cinematic experience. By the 1960s, headsets and goggles became central to VR development.

Metaverse Potential

Companies and investors are ready to spend. And so are consumers. The growth in investment in AI has been huge, from $18 billion in 2016 to $92 billion in 2022 and will exceed $300 billion by 2025.[43] Investment in the metaverse is likely to grow at an even faster pace. Businesses are building infrastructure and crafting virtual worlds. Consumers are purchasing enabling hardware and acquiring digital assets. The fusion of virtual currencies and NFTs with everything metaverse exponentially deepens the potential of this next wave of technology. Innovators and experts expect the technology and immersive experience to improve as companies compete to achieve and commercialize breakthrough products and services.

Potential towers over uncertainty. The metaverse is becoming the most significant growth opportunity across multiple sectors and within many industries. This diversity of potential helps account for the exploding investment that companies around the globe are making. Potential is also driven by the sheer multitude of ways in which the metaverse can be used. The flexibility and the elimination of any discontinuity between imagination and experience introduces a limitless set of possibilities.

Across Economic Sectors

The metaverse will reach all economic sectors (and we'll explore those in future chapters). Goods, services, manufacturing, health care, communications, media, finance, fashion and apparel, and entertainment are some of the areas of impact. Scholarship and the study of the long-term and short-term impact of the metaverse on users along with governance and regulatory issues represent a component of potential and will also contribute to value creation.[44]

Some metaverse business activities will use the VR space to market and sell non-VR products, services, accommodations, and experiences.

Real-world items may direct consumers to the metaverse through QR codes or internet links. As the metaverse integrates into routine operations, new business models and ways of monetizing, marketing, and selling will emerge. It has the potential to revolutionize instruction and training, potentially reshaping all levels of education.

Brand loyalty and customer engagement are already finding a place in the metaverse. Virtual e-commerce platforms are in place, and immersive shopping experiences will expand what businesses currently do with their internet stores. Virtual tourism offers endless options, from museums to beaches to famous sites, demonstrating the metaverse's capacity to reinvent existing venues and introduce new products, services, and activities.[45]

The metaverse will change how people work, how businesses interact with customers, and how companies operate broadly. Marketing, sales, customer service, and support are already exploring new ways to connect.

Choose Firsthand Knowledge

Leaders must quickly familiarize themselves with the metaverse. Exploring this space can provide crucial insights, similar to how early internet adopters like Amazon, Google, eBay, Yahoo, and Netflix gained an edge. Established companies like IBM, Coca-Cola, General Electric, Intel, and Ford also benefited from having an early internet presence.

These success stories should encourage leaders to engage with the metaverse. Platforms like Decentraland and The Sandbox offer opportunities to get acquainted and understand the potential impact.[46] For executives, using VR headsets for a virtual tour can offer valuable firsthand experience.

Understanding the metaverse's impact is linked to creating value. By immersing themselves in this new digital realm, leaders can innovate and envision how their brands, products, and services can thrive. This hands-on approach helps in identifying opportunities for improvement and expansion in the metaverse.

Focus on the Human Component

At the outset, focusing on the human component is crucial and offers a strategic advantage.[47] Balancing the technological potential with creating optimal, healthy user experiences is essential. Leaders who immerse themselves in the metaverse gain valuable insights into the urgency of providing a safe and rewarding experience for consumers. For instance, using MetaMask (a trusted crypto wallet, which allows the user to buy, send, spend, swap, and exchange digital assets) highlights the importance of securing user data, while engaging with platforms like *Fortnite* underscores the need to protect children and teens, offering businesses opportunities to innovate.[48]

The metaverse promises multiple layers of societal change.[49] Unlike the current internet, it exposes more of an individual's biometric and behavioral data, including eye-tracking data. This richer dataset will likely be the focus of future law and policy. Responsible leadership must ensure that products and services are inclusive, accessible, and cause minimal user distress. Businesses that develop practices to support and protect users will be at the forefront of metaverse development.

AI is already a driving force in shaping the metaverse, but addressing algorithmic bias is fundamental to fostering a digital space that treats all users fairly.[50] The way these algorithms are designed today will have long-term implications for economic participation, digital identity, and societal interactions. From reshaping education, health care, and tourism to transforming commerce and entertainment, the metaverse has the potential to enhance everyday life. However, achieving this potential requires thoughtful, intentional design that prioritizes inclusivity and usability.

Businesses engaging with the metaverse at this stage have a unique opportunity to influence its trajectory. The companies that consistently listen to consumer feedback, adapt to evolving user needs, and invest in responsible innovation will not only drive the metaverse's growth but also shape a digital world that enhances human interaction rather than replacing it.

Early Engagement Is Crucial

The metaverse is gaining daily attention but remains a moving target. Businesses and leaders should avoid waiting for it to mature too much before getting involved. Early adopters will face initial challenges but also reap significant rewards, gaining valuable early expertise.

Amazon's evolution from books to a vast array of products, Google's expansion from the most basic of search engines, and Netflix's rise from video rentals to streaming exemplify the benefits of early engagement. Businesses can't predict the exact lessons they'll learn from the metaverse now, but the skills and adaptability gained will be crucial for future success.

Efforts to create VR have spanned centuries, and now that it's here, the blend of ordinary and extraordinary offers vast benefits across sectors like education, medicine, retail, and professional services. By engaging in the early stages of the metaverse, leaders and businesses can effectively manage and respond to both predictable and unforeseen trends and opportunities.

3 The Behavioral Shifts

One thing is certain: no one can predict all the ways in which the metaverse will reshape our current activities or introduce new ones.[1] We cannot easily foresee how the metaverse will alter us. A look at the history of major innovations suggests opportunity, disruption, and upheaval. However, let's think smaller for a moment. The internet has forever changed the way we play games, both together and alone. Take solitaire, for instance. This card game appeared in northern Europe in the late 1700s, with earlier traces found in Asia. The 1990 release of Windows 3.0 introduced users to a digital version of solitaire, and the advent of the World Wide Web in the 1990s widened access to the game. One thing is certain: there will be plenty of solitaire in the metaverse, but it will likely look as different from Windows and internet versions as those versions did from the card-based game.

Just as the internet changed how we played games, accessed news, and paid bills, the metaverse will reshape our behaviors. The ways in which leaders align their businesses with these changes can foster a competitive, healthier model of the company–customer relationship.

Face the Changes

The metaverse is not all entertainment and recreation. Much of what we do routinely now will be done in the metaverse. By reflecting on how the internet rapidly and radically altered our relationships with information, communication, commerce, creativity, activism, and social interaction, we can anticipate how extensively the metaverse will modify who we are.[2] The internet also introduced cyberbullying, trolling, and online addiction, and it still perpetuates hoaxes, urban legends, and misinformation. It disrupted business and behavior.[3]

The metaverse may be a saving influence. If we consider the metaverse akin to a *Star Trek* holodeck, it can be used for education, training, communication, research, simulation, social gatherings, religious rites, therapy, problem-solving, prototyping, and escapism. The routine and the remarkable are already feasible in the metaverse: paying bills, transferring funds, ordering meals, attending meetings. If it all becomes too much, you can attend a basketball game or head to Paris, Vegas, or Seoul. Feeling adventurous? Get virtual plastic surgery, visit Saturn, ride an extinct animal, or live a completely alternate life—complete with a partner and children.

As we immerse ourselves in the metaverse's activities, we will change.[4] The metaverse can sustain old, undesired behaviors or offer options that support better mental health. The outcome, positive or negative, will be significantly driven by inclusive design.

The Challenge Continues

The internet continues to evolve while grappling with challenges in representation, safety, accessibility, and user engagement. Early digital platforms lacked input from a broad and diverse range of innovators and policymakers, leading to gaps in usability and inclusion. While ongoing efforts aim to bridge these gaps, Web 3.0 enters the landscape with the advantage of hindsight—a view that acknowledges how seemingly small design choices can significantly impact engagement, accessibility, and overall user experience. Whether in real life (RL) or virtual reality (VR), design choices can invite users in—or leave them out.

The metaverse's ability to attract, retain, and empower users will be shaped by how well technology integrates diverse needs and perspectives. This requires forward-thinking policies and inclusive design principles that ensure digital tools—whether social platforms, smartphones, or immersive VR environments—are welcoming and adaptable for all. How people engage in learning, marketing, commerce, socializing, and entertainment will be fundamentally redefined in the metaverse. These technologies do not merely respond to human behavior; they shape and guide it. What people can do in the metaverse will be dictated by the structures put in place, and user demand will ultimately shape the priorities of innovation.

Web 3.0 aims to move beyond the limitations of earlier internet iterations, introducing more immersive and intuitive experiences. However, the speed of adoption and the metaverse's long-term viability will depend on the decisions of today's innovators and industry leaders. Engineers, designers, and business strategists must recognize that there is no place for outdated exclusions or biases. They must build future-proofed solutions that anticipate and address user needs—because if they don't, someone else will. Thoughtful leaders must also consider user well-being, ensuring that metaverse engagement fosters positive experiences rather than reinforcing digital fatigue or exclusion.

As artificial intelligence (AI) and machine learning (ML) become increasingly embedded in the metaverse, the risks of algorithmic bias must be actively managed. AI-driven tools influence everything from avatar representation to social interactions and content curation. If unchecked, these systems can unintentionally perpetuate disparities or reinforce preexisting stereotypes.[5] To avoid this, businesses must adopt intentional, data-driven strategies that prioritize accuracy, fairness, and adaptability. A metaverse designed with proactive oversight and responsible innovation will become a space where users feel represented, valued, and in control of their experiences—rather than sidelined by automated processes.

Technology Addiction

It may be difficult to distinguish between necessary utility and addiction when it comes to aspects of digital technology.[6] Technology addiction has been well researched and linked to anxiety and depression, particularly among younger users. Leaders and innovators may not set out to facilitate unhealthy connections between their users and their tech, but the negative outcomes are well documented. Push notifications, infinite scroll, and other features can hold individuals entranced in a state of perpetual engagement. Moving into the metaverse requires proactively including measures that shape user behaviors and support mental health. Addiction within this space is not an unavoidable outcome. Integral to a company's measure of success in the metaverse will be its capacity to care for customers.[7]

Improving Our Well-Being

A thoughtfully designed metaverse can enable and drive activities to improve our physical and mental health.[8] Many popular products for VR environments already offer immersive opportunities for physical exercise, and more are sure to come. This kind of activity combats the depression that can arise as a result of tech addiction. Science has demonstrated that exercise helps with stress management and improves sleep. Smartwatches and many smartphone apps serve these functions. The metaverse can add an immersive dimension and take advantage of more detailed biometric data.

Getting mental and physical health right in the metaverse and working to avoid over-engagement among users has to be a priority of engineers and designers. Customers are constituents of the metaverse; they cannot be "on" 24/7. Some platforms insert messages if a user has been scrolling without really engaging content. In the metaverse, businesses can provide tools and resources to help users track and manage time spent. They can educate users about the risks of addiction and how to avoid them. Businesses can also design products and services that minimize addiction risks and partner with mental health organizations to provide resources to struggling users. Certain age demographic groups such as children are more likely than others to suffer from digital addiction, and keeping this information front of mind can help businesses design the metaverse more responsively and inclusively.[9] Leaders should tread particularly carefully when thinking about the risks of tech addiction among young adults and children.

Healing from Web 2.0

The metaverse offers a unique opportunity to address some of the challenges that emerged from Web 2.0 platforms, particularly social media. Over the past two decades, the internet has both connected and segmented people, creating a digital landscape where interactions can strengthen relationships but also contribute to polarization. Social media platforms, originally intended to enhance communication, have at times reinforced ideological divides, elevating sensational content and limiting exposure to diverse perspectives. Algorithmic decisions, driven by engagement metrics, have often prioritized controversy over constructive dialogue. As the metaverse evolves, there is potential to build digital spaces that encourage more balanced and meaningful interactions, fostering engagement that prioritizes depth and nuance over division.

For the metaverse to truly innovate, it must avoid replicating these digital pitfalls. Leaders, engineers, designers, and content creators now have the responsibility to develop a digital landscape that promotes creativity and dialogue without enabling harmful content that has become ingrained in Web 2.0. The challenge lies in ensuring free expression while designing systems that do not perpetuate division, disinformation, or exclusion. Psychological safety must be at the core of this transformation, requiring thoughtful design and a proactive approach to fostering healthy digital communities.[10]

Those shaping the metaverse's infrastructure must also recognize that every design choice sets a precedent. Will metaverse environments be static and universal, appearing the same to all users? Or will they be customized based on factors like user profiles, preferences, location, education, or ideology? While personalization can enhance experiences, it also carries the risk of reinforcing digital silos—replicating the algorithmic traps of Web 2.0.

However, unlike traditional digital platforms, the metaverse offers new ways to reframe identity and interaction. Many of the assumptions we make about others—based on sensory cues such as voice, appearance, or physical characteristics—can be de-emphasized or even redefined through avatar customization. A user can shape their digital identity in ways that challenge real-world biases.[11] For instance, a woman might choose a male avatar, a Black user might select a white avatar, or someone with limited mobility might experience the metaverse with an avatar designed for full mobility. While this ability to design one's digital presence can bypass surface-level biases, it does not solve the deeper societal issues that perpetuate discrimination in the first place.

Customization also raises important ethical and practical questions. In contexts like virtual home tours, personalization may provide users greater control over their digital presence. But in professional environments—such as job interviews or financial assessments—the possibility of selecting an "advantageous" avatar underscores the persistence of systemic bias. Ideally, the metaverse should cultivate environments where people can represent themselves authentically without fear of discrimination or disadvantage.

At the same time, avatar customization can also be a tool for education and empathy. In professional training or learning environments, users could be assigned avatars that offer insight into different lived experiences—helping individuals understand perspectives outside of their own. This functionality could reshape corporate training, leadership development, and social awareness initiatives, transforming the metaverse into a space that not only reflects but actively improves how people engage across differences. If built with foresight, the metaverse can be more than an escape from the failures of Web 2.0—it can be a corrective force, designing better digital interactions from the start.

Reducing Adverse Behaviors

The internet has become a driver of adverse behaviors, including disruptive and violent disagreements among groups and individuals. Additionally, social media and other websites play significant roles in promoting negative self-directed behaviors, such as poor body image. Can the metaverse reduce these adverse behaviors, especially those impacting teenagers?

The metaverse can be designed to improve long-term behaviors and enhance individual mental health.[12] Businesses have the opportunity to shape the tone of consumer and customer engagement through the content and messages they direct at individuals in these virtual environments. While marketing can influence purchasing choices, the metaverse's potential for personalized well-being solutions goes beyond consumerism. With appropriate consent, data such as biometric readings, emotional inputs, food consumption patterns, and medical information could be integrated to create customized therapeutic strategies and holistic care plans. This approach would prioritize advancing well-being and developing innovative health interventions rather than serving marketing purposes. However, to achieve this, strict compliance with data privacy regulations like HIPAA in the United States is essential to ensure sensitive medical information is securely handled and used solely for patient benefit. By proactively addressing privacy concerns, the metaverse could unlock transformative possibilities for health and well-being.

Taking care of the customer will be a key strategy in the metaverse.[13] Fostering business success and customer well-being are not mutually exclusive. Ethical design sustains the organization and its client base. Using behavioral science to shape positive choices creates wins for both the consumer and the business. These outcomes can be observed and measured through increased customer trust and brand loyalty. Leaders' choices can convey that the company values its connection to the customer and prioritizes their well-being. One way to implement this commitment is to design the metaverse to enable customers to find and assess valid information on topics that impact them.

The Behavioral Shifts

The data collected by VR headsets and haptic devices contain a trove of information about the user. Users' activities and the environments they visit, combined with identifying data and responses to surveys, can enable AI tools to curate information or offer suggestions relevant to a given user. When suggesting information, even if it is for the purpose of promoting a product or service, businesses can provide connections to vetted underlying sources. This practice can help cultivate responsible thinking and begin to heal the damage caused by misinformation.

Can I Trust It?

In an era where the lines between fact and fiction blur with every click, the metaverse stands poised to become the next frontier for information dissemination. In this realm, immersive environments wield more influence than traditional print or video formats. As we've seen during the internet age, particularly throughout the COVID-19 pandemic, misinformation has found fertile ground on social media platforms, leveraging their vast user bases and high engagement. While these platforms have struggled to mitigate the spread of falsehoods, they have also inadvertently benefited from the very misinformation they seek to control.

The transition from the 2D, static experiences of the internet to the 3D, immersive realms of the metaverse presents both a challenge and an opportunity. In the metaverse, users don't merely browse—they inhabit. This deeper immersion means that misinformation could potentially be more persuasive and more enveloping. Yet, within this immersive setting, the metaverse also offers a unique avenue to redefine how information is moderated. AI, central to this new domain, holds the key.[14] With thoughtful training, AI systems can become adept at identifying and removing harmful content, including hate speech and misinformation, while avoiding the amplification pitfalls seen on current platforms.

However, the metaverse's capacity to be a safe space, resistant to the tides of misinformation, hinges significantly on how these AI systems are calibrated and educated by the companies at the helm. As the metaverse evolves, it becomes imperative for those designing and governing these virtual spaces to imbue them with mechanisms that prioritize the well-being and enlightenment of their inhabitants. The lessons learned from the struggles against misinformation in the age of social media must inform a proactive approach in the metaverse—ensuring that as users step into these new worlds, they enter realms where truth is upheld and misinformation finds no refuge.

Customer Well-Being as a Competitive Advantage

Designing for customer well-being can create a new competitive advantage in the metaverse.[15] This new digital realm is expected to reignite social dynamics that are less feasible in Web 2.0. Users may engage deeply and then easily return to RL, or they may struggle to disconnect, showing signs of addiction. Understanding the history of the internet's impact can help businesses plan better for the metaverse era.

In the metaverse, organizations and consumers can coexist as a more focused and healthier whole. The traditional approach of "capturing" and "harvesting" consumers should be replaced with a focus on guiding and informing them. Consumers now help businesses focus on the right projects and introduce ready-to-adopt innovations, while businesses help consumers avoid old pitfalls and choose beneficial behaviors. This shift allows businesses to become more embedded in their relationships with customers and entrusted with significant aspects of their well-being.

Hyper-engaging users in the metaverse can bring immediate benefits in user time and revenue, but this approach is not sustainable. Instead, businesses should build the metaverse to foster well-being, sustain users, and create global understanding. Beneficially reconfiguring behavior as the metaverse develops will be crucial for businesses and their leaders. Given the detrimental behaviors that emerged during the internet's growth, early intervention and proactive inclusive design are essential.

The metaverse offers businesses a closer connection to customers, allowing them to better "see" and gauge customer well-being. While cookies and SEO provide a hazy outline of people on the other end of marketing and transactions, metaverse-accessible biometric data and interaction records offer a more substantial view. Businesses can choose how to use these new layers of data and insight. Misuse or abuse of this data will undermine trust and the inclusive potential of the metaverse.[16]

By focusing on well-being and ethical design, businesses can create a sustainable and inclusive metaverse that benefits both users and companies.

Extremism Won't Disappear

The metaverse, like previous internet platforms, will not be immune to the risks of indoctrination and radicalization.[17] Online spaces have long facilitated exposure to extreme ideologies, often leveraging the same tools used for education and engagement, such as posts, videos, and live discussions. The immersive nature of the metaverse introduces new complexities—while it offers powerful applications for training in areas like medicine and engineering, it also carries the risk of being exploited for harmful purposes, including unauthorized simulations or extremist recruitment tactics. Proactive and responsible leadership in the metaverse requires a broad, strategic approach that goes beyond algorithmic monitoring. Strategic partnerships between businesses, educators, and security experts will be crucial in shaping an environment that fosters constructive engagement rather than unintentionally amplifying divisive or dangerous content. Companies developing metaverse technologies must consider not only how their platforms function but also how they can create meaningful safeguards against misuse.

While data-driven insights can help identify risks, overly aggressive surveillance and personal data tracking may raise ethical concerns and erode user trust. Instead, businesses should prioritize designing secure, engaging, and well-moderated virtual spaces. By working with experts in digital safety, behavioral science, and education, leaders can help create a metaverse that expands opportunities for learning and collaboration while reducing the potential for exploitation and harm.

Avatars and Body Image

Social media's impact on negative body image could be significantly amplified in VR.[18] The idealized bodies prevalent on social media have intensified body image obsessions, particularly among teens. In a culture grappling with obesity and nutrition issues, young people are especially vulnerable to unhealthy choices in their quest for thinness.

As various slices of the metaverse are rolled out by decentralized service providers, designers, content creators, and innovators, leaders must remain aware of the potential for harm. Choosing an avatar is a complex task when it comes to body image. Selecting an avatar that accurately represents one's body can lead to body shaming, while choosing one that looks different from one's real self can cause private, internalized shame.

Leaders must proactively address these challenges by promoting diverse and realistic body representations in the metaverse and providing support and resources for positive body image. Integrating educational content and promoting body positivity can help mitigate the potential negative impacts on users, especially teens.

How Much to Worry

Concerns about potential harm in the metaverse are valid and not overly extreme.[19] The alarm people feel as the metaverse becomes inevitable is proportionate to the significant risks involved. Many things could go wrong, but there are strategies to mitigate these risks.

While solving technical challenges, offering the best environments, and realizing substantial earnings are important, priorities should focus on users' betterment and well-being. Behavioral science and behavioral economics can help organizations reconcile seemingly conflicting goals. Integrating psychological wellness resources into the metaverse is essential, and insights from behavioral scientists should guide its development.

For example, virtual reality can simulate social scenarios to teach empathy and understanding, reducing prejudice and fostering inclusivity. It won't be enough to simply encourage self-care or healthy choices; the way messages are crafted and delivered will determine their effectiveness. Behavioral science provides insights and tools that empower users to maintain personal sovereignty while making decisions that enhance their well-being.

Intentional Design

In crafting the next frontier of digital interaction, intentional design principles are essential for ensuring the metaverse evolves beyond entertainment into a dynamic, accessible, and engaging space for all. This design ethos envisions a metaverse that seamlessly integrates customer–business relationships in Web 3.0 environments, fostering a balanced, user-friendly, and innovative ecosystem. It rejects the notion that inclusivity and engagement are at odds—a well-designed metaverse can offer both excitement and accessibility without compromise.

Strategic design choices for the metaverse require a proactive approach from leaders, innovators, and early adopters, working alongside lawmakers and industry stakeholders to establish best practices from the ground up.[20] It's about identifying and eliminating barriers that could limit participation, ensuring that virtual spaces are designed to be engaging and accessible across physical abilities, socioeconomic backgrounds, and different levels of digital fluency. For example, expanding user interfaces beyond traditional keyboard and touch inputs—such as voice commands or gesture-based controls—can allow broader participation, particularly for individuals with disabilities.

Furthermore, thoughtful metaverse design has the potential to enhance real-world relationships, strengthening social and family connections while creating new ways for businesses to tailor their products and services. This includes the evolution of AI-driven interactions that can adapt to diverse user needs and foster positive, meaningful engagement.

Leaders and developers have a unique opportunity to shape this emerging digital universe. Prioritizing user-first, inclusive design from the outset ensures the metaverse becomes a vibrant, multidimensional space where technology serves to connect, empower, and elevate human experiences. This approach positions the metaverse not just as a technological achievement but as a reflection of our commitment to creating a digital space where all users can contribute, participate, and thrive.

The Metaverse Will Reshape Us

We can safely predict that the metaverse will profoundly reshape human experiences.[21] Businesses need to anticipate these changes and actively shape the metaverse in ways that enhance user engagement, support societal well-being, and unlock new opportunities for consumers. The immersive nature of the metaverse will redefine how people communicate, shop, play, work, learn, and interact, fundamentally altering our relationship with technology and each other.[22] Skepticism toward new technology is nothing new. Just as early critics doubted the practicality of personal computers, mobile phones, and social media, Web 3.0 faces its own share of naysayers. However, historical patterns suggest that initial hesitation often gives way to widespread adoption. Will the metaverse make us better? Possibly. Worse? Perhaps. But without question, it will make us different.

The internet reshaped many simple joys of life, like social interactions in person or over the phone. However, the immersive nature of the metaverse can offer a return to a social component in many aspects of life. The metaverse can level the global playing field, offering immersive connections and educational opportunities to isolated or underserved communities. Access to VR labs and training facilities can provide real, marketable skills. Poor connectivity or costly products could create a new digital divide, but where beneficial factors converge, significant educational and economic changes are feasible.

There is no shortage of projects that historically didn't turn out as planned.[23] Predicting negative outcomes is not fully within our capacity, but recognizing and anticipating negative user responses or community impacts can save businesses time and money. The internet's history offers lessons; anticipating the metaverse's impact can guide better planning and inclusivity. A healthy user base is more economically viable, and VR can be designed to support mental, social, and physical wellness among users.

Leaders do not have the luxury of waiting for studies to confirm the metaverse's addictive elements. The metaverse has already proven addictive to some. Obsession with VR and excessive time spent gaming may become larger issues than excessive screen time. Social media's engagement has often led to toxic behavior and unhealthy social media content. The metaverse must avoid replicating or intensifying these ill effects. Prioritizing user well-being in a comprehensive and proactive way is essential.

Real-world problems will follow users into the metaverse. As with social media, both positive and negative behaviors will be amplified. While social networks have deepened personal connections, they have also enabled troubling trends, such as coordinated flash mobs and, in some cases, organized retail theft. The metaverse's ability to host unlimited simultaneous users and complex interactions will magnify both opportunities and risks. Psychological disinhibition—encouraged by anonymity and immersive environments—will foster both creativity and potentially harmful behavior.[24]

Regulatory and safety frameworks must evolve alongside these technological shifts. Anonymity and virtual interactions create new challenges for law enforcement, legal systems, and platform governance. Policymakers and business leaders must proactively define regulations, security protocols, and ethical standards to deter exploitation and protect users. Eventually, digital law enforcement or regulatory agencies may emerge within the metaverse, requiring collaboration between metaverse innovators and real-world authorities to ensure public safety.

The metaverse's ability to drive both progress and pitfalls calls for deliberate action. Thoughtful planning, user-first design, and proactive policies will determine whether this new digital landscape becomes a force for growth, collaboration, and opportunity—or a repetition of past mistakes on a larger scale.

II A Blueprint for a Thriving Metaverse

4 Unlocking Opportunity

The metaverse has the potential to expand participation beyond traditional digital spaces. In virtual reality (VR), individuals with mobility challenges or physical disabilities may engage in experiences previously inaccessible to them. The metaverse also provides opportunities for people to connect, collaborate, and explore freely, regardless of real-world limitations. However, challenges remain—incidents of harassment, economic barriers, and technological limitations could hinder the metaverse's ability to be truly open to all users. The high cost of headsets and advanced computing requirements may limit access, particularly for individuals from lower-income backgrounds or regions with poor connectivity. Whether the metaverse bridges or widens inequalities will depend on how it is designed and governed. Key factors include affordable technology, safe virtual environments, adaptable controls, and user education.

Inclusive innovation has been spotlighted by international bodies such as the UN for its role in confronting grand challenges like those set forth in the Sustainable Development Goals (SDGs) and Millennium Development Goals (MDGs), including significant efforts toward reducing gender, racial, and socioeconomic gaps.[1] Inclusive innovations aimed at societal betterment have been prioritized in policymaking and leadership agendas since they offer pathways to circumvent the boundaries of traditional welfare systems.[2] Entities such as the European Union (EU) and the World Economic Forum (WEF), as well as large companies, advocate for various inclusive innovations.[3] A particular focus is placed on combating gender, racial, and social exclusion resulting from technological innovations as broadening access to technology benefits not only individuals but entire economies.

Unlocking Growth in the Workforce

While technology has the potential to increase access and drive new economic models, disparities in leadership, employment, and funding continue to influence the design and accessibility of emerging digital spaces, including the metaverse. In the US, the tech sector remains unbalanced in terms of workforce representation, particularly in leadership roles. Data from the Equal Employment Opportunity Commission (EEOC), Society for Human Resource Management (SHRM), and McKinsey & Company highlight these disparities.[4]

Black workers make up only 7.4 percent of the high-tech workforce and 5.7 percent of high-tech managers, despite comprising 12.6 percent of the total US workforce, reflecting a slight increase from 2005 (6%). Hispanic workers are similarly underrepresented, making up 9.9 percent of the high-tech workforce and just 8.1 percent of high-tech managers, while they represent nearly 18.5 percent of the overall US workforce. Asian workers, while comprising 18.1 percent of the high-tech workforce, make up only 15.3 percent of high-tech managers and represent just 6.6 percent of the total US workforce.[5]

Women remain critically underrepresented in the tech industry, accounting for just 22.6 percent of the high-tech workforce—significantly less than their 47.3 percent representation in the overall US workforce. Moreover, women represent only 19.4 percent of higher-paying tech jobs within the high-tech sector. The concentration of younger workers in tech is also notable, with 40.8 percent of the high-tech workforce aged 25 to 39, compared to just 33.1 percent in the overall workforce.[6]

While the racial and ethnic diversity of the high-tech sector (37.4% employees of color) has improved, it still lags behind the total US workforce (41.6%). Women in high-tech (31.9%) fare slightly better than in the broader tech industry, but their representation still falls short of their presence in the overall US workforce.

Beyond representation, workplace challenges persist. Retaliation is the most common discrimination-related charge in the tech sector, followed by age and pay discrimination, which are more frequent in tech than in other sectors.[7]

Despite growing recognition of the business advantages of a varied workforce, data from major tech companies reflects ongoing disparities in representation. In 2022, Microsoft's tech workforce was 47 percent white and 38 percent Asian, with only 9 percent Hispanic or Latino, 3 percent Black or African American, 0.2 percent American Indian or Alaska Native, 0.2 percent Native Hawaiian or Other Pacific Islander, and 2.6 percent identifying as two or more races. Men comprised 78 percent of this workforce, with women making up 22 percent.[8] At Meta, tech employees were 51 percent white, 18 percent Hispanic or Latino, 17 percent Asian, 5 percent Black or African American, 0.2 percent American Indian or Alaska Native, 0.2 percent Native Hawaiian or Other Pacific Islander, and 8.4 percent identifying as two or more races. Men were 68 percent of Meta's tech workforce, and women were 33 percent.[9] The figures at Apple and Google show similar trends, highlighting the challenge of gathering a significant collection of diverse voices.[10] Only Microsoft publishes data for LGBTQ+ and people with disabilities within their tech staff, at 10 percent and 4 percent respectively in 2022. Meta and Google report LGBTQ+ representation at 10 percent and 7 percent, respectively.

These workforce patterns can directly influence how metaverse platforms are designed, who they serve, and how they evolve. Without broader representation in leadership and development roles, critical design considerations—such as accessibility, user experience, and market reach—may be shaped by a limited range of perspectives.[11] Companies that prioritize adaptive design, user engagement, and diverse talent pipelines will be better positioned to build more scalable, widely adopted metaverse solutions.

How Biases Can Impact Innovation

Biases—both structural and cognitive—can shape decision-making and limit opportunities for participation, innovation, and economic expansion.[12] Biases exist at multiple levels—some are unconscious, others structural, and many are deeply ingrained in decision-making processes. These biases often go unnoticed but can significantly shape product development, user experiences, and engagement. Recognizing and mitigating bias is not just about compliance or ethics—it is essential for building an adaptive, user-friendly, and widely adopted metaverse. Developing a metaverse that minimizes unintended bias requires business leaders, designers, and innovators to actively seek diverse perspectives. This may involve gathering broad user feedback, conducting independent audits, or implementing structured review processes to detect and correct blind spots in product development. Bias is not always intentional, but failing to address it can result in limited user adoption, reputational risks, and even legal challenges.

Designing the metaverse for broad engagement requires deliberate efforts from leaders, innovators, and early adopters to collaborate with industry stakeholders and policymakers. This collaboration ensures that the foundational layers of the metaverse prioritize user well-being, adaptability, and responsible engagement. A well-designed digital ecosystem must proactively address barriers to participation rather than retrofitting solutions later. Creating environments that are intuitive, flexible, and user-friendly will allow for broader adoption across diverse populations. Expanding interaction methods beyond traditional inputs—such as gesture recognition, voice controls, and adaptive technologies—can enhance accessibility. Ensuring compatibility across various devices and network conditions can prevent economic barriers from limiting participation. Customizable user experiences, including avatar and interface personalization, can help users tailor engagement to their individual needs and comfort levels. By integrating these principles from the outset, businesses can develop a metaverse that is not only engaging and immersive but also future-proofed for a global audience.

Moreover, intentional design positions the metaverse as a connective fabric that strengthens family and social ties while offering businesses unprecedented opportunities to tailor communications and services. This shift extends to empathetic artificial intelligence (AI), which can adapt to the nuanced needs of diverse user groups and foster virtual spaces that encourage positive, meaningful interactions. AI, particularly large language models (LLMs) and natural language processing (NLP), plays a crucial role in enhancing accessibility by enabling real-time language translation, voice interaction, and personalized user experiences. However, the risks of embedding biases within AI systems are significant. LLMs trained on biased data can inadvertently reinforce exclusion, limiting participation in diverse virtual environments. To mitigate this, leaders and developers must implement proactive measures such as bias detection and correction tools to ensure that the metaverse serves as an adaptive and widely accessible digital space.[13]

Leaders must recognize their influence in shaping this evolving digital landscape. By prioritizing user-centered design from the outset, the metaverse can grow into a dynamic, multifaceted realm where innovation flourishes alongside adaptability and engagement. The goal is to create a digital ecosystem where every individual finds a place, a voice, and meaningful participation, ensuring the metaverse evolves as a technological breakthrough and a model for inclusive, human-centric development.

Decision-Making and Cognitive Biases

Navigating the complexities of metaverse development requires an awareness of the biases that influence decision-making. Two primary types of bias frequently shape workplace environments and innovation processes.[14] The first type includes cognitive biases—errors in logic and flawed group dynamics—such as confirmation bias and the sunk-cost fallacy, both of which are well documented in psychology and behavioral economics.[15] The second, more systemic type of bias involves prejudices against specific groups of people, which can create barriers to participation and limit diverse contributions in digital spaces.[16]

Biases can hinder business success by leading to poor decision-making, workplace conflict, and stalled innovation, ultimately impacting revenue and market competitiveness.[17] While organizations rarely intentionally embed bias into their structures, these patterns often emerge as shortcuts in decision-making—appearing efficient but ultimately limiting creativity and growth. Leaders across industries must recognize and mitigate both forms of bias to ensure their metaverse strategies remain adaptive, forward-thinking, and open to a broad range of perspectives.

Errors in Logic or Flawed Group Thinking

Biases from errors in logic or flawed group thinking are cognitive errors that can lead to irrational or inaccurate judgments.[18] They arise from personal biases, the influence of groups, and heuristics (mental shortcuts) used to process information. Common biases include confirmation bias, overconfidence bias, and the sunk-cost fallacy, which are pervasive beyond the workplace.

Confirmation bias inclines observers to focus on data that align with their existing beliefs, creating an echo chamber where new ideas struggle to arise. If your conclusions always match your expectations, you may have slipped into confirmation bias. To counteract this, listen to diverse perspectives during team discussions and be open to considering new information that contradicts your existing beliefs. Encourage disagreement and cultivate diverse teams to foster innovation and prevent future product failures.

Overconfidence bias—an overestimation of one's abilities and aptitudes—also plays an undermining role in innovation efforts.[19] In startups, passion and excitement, while essential, can lead to overconfidence. It's important during product development to remain realistic, ensure objectivity, and consider new ideas. Transparency and feedback processes, like the premortem framework, can help identify and mitigate overconfidence bias.

The *sunk-cost fallacy* can be just as detrimental to business operations and innovation as the aforementioned biases. After an organization has poured substantial and potentially irrecoverable investments of time, talent, and capital into a project, withdrawing these resources and terminating efforts can occasionally be the best strategic choice.[20] Innovation requires knowing when to cut losses. If a project yields the same failed results despite different approaches, it's time to reconsider. Be realistic about timeframes and recognize when a project is failing.

Understanding and addressing bias is essential to avoiding critical missteps in product design and user experience.[21] When decisions proceed unchecked or lack input from a broad range of perspectives, they can lead to overlooked needs, flawed assumptions, and unintended consequences. For example, many so-called "women's" products differ only in size or color rather than function, failing to account for gender-specific needs. This oversight is not just an inconvenience—it can have real-world safety implications. In the auto industry, for instance, women are 73 percent more likely to be injured in a crash due to historical testing biases that primarily used male crash-test dummies.[22]

Bias also affects the creative process itself. When innovation takes place in an echo chamber—where the same perspectives are reinforced without challenge—companies risk stagnation. The businesses that thrive in the metaverse will be those that invite, empower, and integrate diverse innovators into their development process. By actively seeking out and incorporating broader perspectives, companies can create safer, more functional, and widely adopted products. Removing barriers to innovation and decision-making not only enhances user experiences but also provides a competitive advantage in the rapidly evolving metaverse landscape.

How Bias Shapes Perception

Cognitive biases can significantly influence decision-making, particularly in how individuals and groups are perceived in professional and innovation-driven environments.[23] One of the most prevalent biases, confirmation bias, reinforces preexisting beliefs by filtering out contradictory information. When applied to groups, this bias can lead to stereotyping, limiting opportunities for individuals who do not fit traditional expectations. For example, in the tech industry, individuals who do not align with the stereotypical profile of an "innovator" may hesitate to contribute ideas, fearing their insights will be dismissed. This can stifle creativity, collaboration, and the diversity of thought needed for breakthrough innovation.

Other biases, such as the halo and horn effects, also shape workplace perceptions.[24] The *halo effect* causes us to generalize one positive trait to assume overall excellence. For example, if someone is good at coding, they might be perceived as good at project management without any evidence. Conversely, the *horn effect* makes us generalize one negative trait, leading to the assumption that someone who made a mistake is incompetent overall. These biases are prevalent in hiring, promotions, and team selections, potentially sidelining talented individuals who could drive innovation and growth. The implications of who is qualified for which tasks and projects are obvious, as are the potential outcomes for growth, new ideas, and innovation.

These biases can intersect with other social biases, such as those based on gender, race, or nationality. The halo effect might lead us to attribute positive qualities to someone based on one good trait and link these to stereotypes. For example, a competent male leader might be assumed to be better at handling stress than his female counterpart without any factual basis. On the other hand, the horn effect might lead us to link negative traits similarly, further entrenching stereotypes.

Fundamental attribution error refers to our tendency to attribute others' actions to their character while ignoring situational factors.[25] For example, if an employee misses a deadline, it might be attributed to laziness rather than considering they might be dealing with a personal crisis. This bias can lead to unfair assessments and decisions that overlook external factors affecting performance.

In-group bias occurs when we favor those within our social group, extending favoritism and opportunities to them while discriminating against out-group members. This can manifest in the workplace through preferential treatment, overlooking mistakes, and unfairly extending opportunities to in-group members while excluding or mistreating out-group members. Such biases can hinder the potential contributions of out-group members and limit diverse contributions that could drive innovation. Leaders can mitigate this bias by fostering structured social interactions between in-group and out-group members, which can help reduce barriers and foster a more inclusive and high-performing environment.

Understanding and addressing these biases is essential for leaders who aim to foster accessible innovation for all. By recognizing and mitigating these biases, organizations can broaden perspectives, strengthen decision-making, and cultivate environments where all employees feel valued and empowered to contribute.

A Tool for Overcoming Biases?

The metaverse offers a unique platform for self-expression, allowing individuals to create diverse avatars that enhance their ability to express themselves and reduce social stigmas. For those constrained by societal pressures in the real world, the virtual space can serve as a sanctuary to explore different aspects of their identity without fear of judgment.[26] Users often report that avatars perceived as more attractive than their real selves encourage more outgoing and adventurous behaviors. Research suggests that VR and the metaverse can positively shape users' self-concepts, promoting self-acceptance and focusing on achievable personal change rather than unrealistic beauty standards.[27]

The variety of avatar customization options in the metaverse enables users to portray their genuine selves, fostering individual expression and identity. This freedom enhances personal and collective well-being and boosts self-esteem.[28] Studies such as those by Friends with Holograms reveal that using anonymized avatars in professional settings can improve collaboration and engagement, especially among lower-level employees.[29] Additionally, the metaverse and VR serve as powerful platforms for empathy enhancement, providing valuable tools for understanding and adopting alternative perspectives.

An Unexpected Challenge

A well-designed metaverse can become a powerful global economic engine, offering earning opportunities from global tech giants to remote individuals in the creator economy. However, a significant barrier exists: around 1 billion people globally lack legal identification.[30] Poverty, conflict, displacement, lack of resources, remote location, and discrimination contribute to the absence of identifying documents and credentials, inhibiting individuals' ability to engage in essential societal functions like opening a bank account, obtaining a loan, or voting. This issue is especially prevalent in low-income nations, with roughly 45 percent of women and 28 percent of men lacking legal ID.[31] People without legal identification face difficulties accessing essential services, are at higher risk of exploitation, and may be socially and economically excluded.

The UN has prioritized ensuring universal legal identity by 2030 as part of its Sustainable Development Goals (SDGs). Implementing digital identities in the metaverse requires careful consideration by regulators and organizations in both the public and private sectors to ensure that those without a legal ID are not barred from securing a digital one. The discussion must also address data privacy compromises and potential complexities linked to cybercrime, financial crime, and other abuses, such as a single user creating multiple IDs. Establishing an ecosystem of reliable digital identity issuers will be crucial. This will prevent individuals from being excluded from new opportunities in education, finance, and social interaction as a result of an inability to procure a digital ID, while also instituting trust levels to minimize fraud.

The metaverse stands out as a novel method for providing identities to those currently absent from ID systems. It may pave the way toward inclusive access to various services and opportunities, bridging the gap for those without traditional forms of identification while maintaining a balance to avoid potential exploitation and misuse. This duality of inclusivity and security in the metaverse's digital identity realm will be paramount in shaping a virtual world that is both accessible and safe.[32]

Balancing Engagement and Safety

Currently, the majority of users are males aged 18–34, with approximately 73 percent identifying as white, reflecting demographics similar to the traditional gaming industry.[33] However, as virtual spaces evolve, ensuring a welcoming and secure environment for a broader user base is critical. Issues such as virtual harassment and safety concerns—particularly for female users—have already surfaced on platforms like Meta's Horizon Worlds. These incidents highlight the urgent need for better security protocols, content moderation, and user protections.

Developers are swiftly innovating to enhance user safety with measures such as designated "safe zones," where avatars can interact without the risk of harassment or the influence of biased algorithms.[34] For instance, initiatives like the "Harmony Hub" create neutral spaces where avatars are anonymized, and interactions are valued based on contributions rather than avatar characteristics. Additionally, "defensive avatar maneuvers" are being developed—a suite of tools and gestures that avatars can use to instantly report misconduct, block unwanted interactions, or even temporarily become invisible to defuse potential conflicts.

These solutions are part of a concerted effort to establish a metaverse that is accessible and safe for all.[35] However, broader and more proactive strategies will be required. Metaverse platforms must go beyond technical fixes and engage in collaborative policymaking, stronger community governance, and enhanced user protections. Ensuring a metaverse that is both engaging and secure will require a multilayered approach, combining ethical design principles, AI-powered moderation, and community-driven safety measures.

What Leaders Can Do

Many default structures and processes in businesses are designed to support specific kinds of people, leading to imbalances in whose ideas are heard. Default, however, does not mean unplanned or haphazard.[36] Adjusting these structures to allow innovations from diverse organization members can move innovation from a small inner circle to a wider arena. Organizations that fail to achieve this kind of inclusion will find themselves at a serious disadvantage.

So, what kinds of biases are clogging your workplace processes? Is your organization suffering from inherent biases in the system or errors in logic and flawed group thinking? Recognizing and neutralizing these barriers are important leadership responsibilities. Changes in thinking and decision-making typically flow from leadership downward through an organization and rarely originate from lower levels moving upward. Leaders set the tone, model desired behaviors, and reset efforts when the drive to demonstrate change wears off. Changed perceptions and behaviors have a ripple effect. Consistent communication about inclusion can change the quantity and quality of innovative output.

A thriving metaverse only happens when all innovators within an organization are engaged. Cognitive biases such as confirmation bias can subtly justify exclusion or even encourage self-exclusion, discouraging individuals from contributing ideas. In the innovation sector, women and underrepresented groups receive disproportionately fewer patents.[37] Companies can shift this reality by cultivating a culture where all employees are recognized as potential inventors, equipping them with the tools, education, and encouragement needed to contribute meaningfully. Gathering data, refining evaluation processes, and establishing clear, fair innovation pipelines will help level the playing field and increase the number of patents and breakthroughs from a wider range of contributors. Leadership recognizes that talent is distributed equally, even if it manifests differently across individuals. Successful transformation requires adaptability, ensuring that different employees' needs and contributions are acknowledged and supported. Organizational strategies should account for how self-perception affects participation in innovation and work to remove barriers that discourage certain groups from fully engaging.

Regardless of a company's size, systematic efforts to expand participation will drive success. Continuous feedback loops, data-driven insights, and structured evaluation processes help leaders course-correct when necessary. Optimizing human resources is not just a moral imperative—it is a competitive advantage. Businesses that prioritize diverse contributions and foster a customer-responsive, innovation-driven metaverse will be best positioned to lead in this new digital era. Rather than simply reacting to past biases, companies must proactively design systems that unlock opportunities for all employees. Leaders who recognize this imperative will ensure that innovation thrives and that their organizations are prepared for the rapidly evolving metaverse landscape.

5 Charting the Path

Why is it essential to develop innovation strategies that reach a broad audience? Expansive innovation ensures more users can engage with and benefit from advancements. Products, services, and processes designed with a wide range of users in mind better address varied needs and challenges, increasing their chances of success in a global, evolving market. Key elements of effective innovation include diverse teams, user-centered design, and open collaboration.

Designing with accessibility and adaptability in mind provides a competitive advantage, as products and services that accommodate different user needs tend to be more intuitive and widely adopted. It also supports economic growth by expanding market reach and increasing usability across demographics. The innovation team should reflect the range of end users, incorporating different perspectives, skills, and lived experiences. This approach helps identify challenges early and integrate effective solutions.

User-driven development is critical. Businesses that actively engage diverse consumer groups throughout design, testing, and refinement produce more functional and responsive products. Collaboration—through partnerships, open-source development, and hackathons—also fuels creative problem-solving and innovation that resonates with a larger audience.

Real-world examples illustrate how broad-focused innovation leads to impactful solutions. The curb ramp, for instance, became widely adopted after advocacy efforts led by Ed Roberts, a disability rights activist. Originally designed for mobility accessibility, curb ramps also benefit parents with strollers, travelers with luggage, and delivery workers. Similarly, smart hearing aids use artificial intelligence (AI) to filter background noise, enhancing communication for people with hearing loss and those in noisy environments. The braille smartwatch allows people who are blind to access digital information and communicate, expanding usability in everyday life. These innovations emerged because designers prioritized usability and accessibility for a broad range of individuals. Conversely, overlooking adaptability in innovation can lead to product failures and unintended consequences. Some facial recognition software struggles to accurately identify users across different racial and gender groups, resulting in misidentifications and technological inefficiencies.[1] Self-checkout kiosks, which now account for nearly 40 percent of all US checkout transactions, pose challenges for people with visual impairments and mobility limitations, leading to lawsuits over accessibility concerns.[2] Additionally, AI-driven voice assistants and reading technologies, trained predominantly on American English datasets, can create barriers for nonnative speakers and reinforce biased narratives about intelligence and communication.[3]

Inclusive innovation must be intentional and systematically embedded in processes.[4] Given the metaverse's newness, businesses must take early steps, even if mistakes are made, to learn and adapt. Leaders need to overcome paralysis, make strategic choices, and formulate strategies that will allow flexibility.

The metaverse offers no guarantees of success, so businesses must communicate their strategy across the workforce, reflect along the way, and remain adaptable to avoid pitfalls and ensure the highest likelihood of success for their products and services.

Behavioral Science and Innovation

Behavioral science partners well with the process of creating new products and services because it offers insights into how people think and form choices.[5] It can enhance innovation at all stages, from ideation to implementation. On the front end, behavioral science can motivate a company's innovators and foster an organizational culture that supports innovation. It can also help identify barriers to participation and engagement, ensuring that all contributors have a voice in the creative process. By addressing low engagement among certain groups, businesses can strengthen collaboration and generate more diverse ideas.[6]

In the middle stages, behavioral science can help teams inclusively innovate products and services that address the gaps and needs identified by diverse team members. At the market stage, it can help businesses

Charting the Path

anticipate how different consumer populations will respond to the look, feel, and function of their products.

Engaging behavioral science allows businesses to better understand user needs and motivations, design user-friendly products and services, and develop effective marketing and communication strategies.[7] It can also be applied to designing and testing prototypes, launching new products and services, and measuring their impact.

Companies like Google, Amazon, and Apple leverage behavioral science in various ways. Google uses it to design its search engine and email platform and to improve marketing and advertising.[8] Amazon employs it to recommend products, enhance the checkout experience, design warehouses, and manage its workforce.[9] Apple uses behavioral science in designing its iPhones and iPads.[10]

Behavioral science has earned its place in the workplace, with many large businesses and government entities maintaining in-house behavioral insight teams. While large businesses can recover from product failures, small businesses often cannot, and they too would benefit from behavioral science insights. Examining the history of innovation through a behavioral science lens can significantly inform product design and user interface. Although innovation processes differ from business to business and technology to technology, behavioral science—being human-focused—offers common, stable elements.

What Behavioral Science Can Offer

Behavioral science can offer insights into what might work and what might meet resistance on both sides of the product equation: innovation by businesses and adoption by consumers. It can help engineers, designers, and inventors facilitate technological breakthroughs, and it holds promise as a tool for navigating consumer reluctance and market resistance. Even if the metaverse represents an ideal opportunity for the right business at the right time, success requires more than that. Businesses must make good decisions about the unique technology required for the metaverse, and behavioral science can help them do so.

The allure of new, innovative, or disruptive technologies may attract investors, but allure often leads to disappointment. Businesses pour money into innovation and product development, but returns may fail to show.

Traditional market studies involve polling or consulting potential consumers about products or services, but consumers do not always know what they want. Understanding the factors that influence behaviors and choices can prevent businesses from squandering capital and opportunities.

In the case of the metaverse, relying on traditional processes can produce ineffective strategies. Manufacturing, marketing, offering, and waiting—combined with research and planning—have paved the way to success for many products and services. However, business leaders who once relied successfully on their instincts might find themselves in new and unfamiliar waters when creating products for the metaverse. Behavioral science offers more accurate tools for testing the potential of new products and services.[11]

Combining a history of behavioral responses to successful and unsuccessful products with reliably predictable social and emotional dynamics removes some of the guessing and trepidation that decision-makers might experience when everything is so new. Behavioral mapping, for example, provides a close analysis of consumer decision-making and can help a business identify opportunities and barriers in consumer needs and interest in a new product or service. Examining the chain of decisions that lead to a purchase removes the reliance on business intuition. With insights from behavioral mapping, innovators can design products that shape consumer decisions and lead to adoption.

Innovation Pitfalls

While consumer demand drives new technology, innovation must also nudge consumers in new directions.[12] This power rests with engineers, designers, and decision-makers. However, along the continuum from invention to consumer adoption, these individuals face various traps and opportunities for failure. In the last chapter we discussed some pitfalls that innovators might fall into. Psychological ownership can wed individuals too closely to their products, making it difficult to make essential decisions. Clinging to unpromising strategies or products, justified by the sunk-cost fallacy, can derail progress. The echo chamber effect, where a group recirculates the same ideas, can blind an organization to other possibilities. Biases such as overconfidence, which leads innovators to overestimate their product's value and underestimate barriers, can lead to failure.[13] The planning fallacy, where the time, cost, and effort needed for project completion are

Charting the Path

underestimated, can further complicate innovation. Confirmation bias, where leaders selectively find information supporting their beliefs, can also lead to failure.

Successful innovation in the metaverse will demand nimble thinking and an ability to accurately assess project potential and a willingness to abandon projects when all signals suggest doing so. Behavioral scientists can help organizations recognize and avoid common pitfalls, making a readiness to let go of what cannot work a hallmark of highly successful players.[14]

Pursuing Accessible Innovation

Consumers want products that reflect their diversity, and the metaverse must meet this demand. Seeing themselves and their experiences in the products and services they use fosters a sense of belonging, making users feel valued and increasing customer loyalty and engagement. The metaverse should represent various aspects of identity, including race, ethnicity, gender, sexual orientation, disability, and body type.

Consumers increasingly seek authentic products and services, supporting brands that share their values. This expectation is particularly strong among younger generations, in particular, who want brands to align with their values and promote equity in meaningful ways. In the metaverse, consumers want avatars that reflect their appearance and identity accurately. They also want culturally relevant products and services, such as accurate skin tones and natural hair textures, religious attire (e.g., hijabs), and assistive devices for users with disabilities.

For businesses, prioritizing adaptability and representation in metaverse development is not just a moral imperative—it is a strategic necessity. Companies that offer multilingual services, collaborate with diverse creators, and integrate inclusive design principles will attract broader and more engaged consumer bases. A gaming company with expansive avatar customization options will appeal to a wider audience, just as a metaverse marketplace that features diverse entrepreneurs and brands will create greater market appeal.

Despite the growing consumer demand for representation in innovation, a persistent gap exists. Statistics from the World Intellectual Property Organization and the US Patent and Trademark Office show that women and minority inventors remain underrepresented. Women are listed on only 23 percent of international patent applications, contributing to just 10 percent of total patents worldwide.[15] Similarly, only 12.8 percent of patent-holders in the US are women inventors, and only 1 percent each are Black and Hispanic inventors.[16] To address these gaps, initiatives like Suzanne Harrison's Diversity Pledge, which unites over 50 key companies, aim to understand and mitigate barriers for underrepresented inventors. The Diversity Pledge asks to publicly share data on the percentage of inventors or creators who are women or from underrepresented backgrounds and the percentage of such individuals who are represented in the company's patent portfolio.[17]

Expanding access to innovation in the metaverse offers businesses a strategic edge.[18] Leadership plays a crucial role in fostering environments where all employees see themselves as inventors and innovators.[19] This involves clear communication of the company's mission and each employee's role within it. When underrepresented members are inclusively engaged, they develop expertise, become more vocal, and contribute more significantly to the company's innovation efforts.

Safe to Fail

Making it safe to fail is critically important for businesses engaged in creating hardware, software, services, and content for the metaverse. Innovating and developing new products and services requires extensive risk and experimentation, and fear of failure can stall a company's progress. Psychological safety is key to staying ahead of the curve. If innovators cannot fail quickly and learn from their mistakes, their innovations (and companies) will become obsolete.

All leaders know about the concept of safe-to-fail, and many want to practice it. Some get it right, but few sustain it. Behavioral economists advocate for making it safe to fail, supported by numerous case studies. Encouraging failure may seem counterintuitive, but it is the art of encouraging all possibilities and unleashing innovation.

In the metaverse realm, where augmented reality (AR), virtual reality (VR), and mixed reality (MR) are reshaping reality, the "fail fast, fail often" philosophy is particularly relevant. The stakes are uniquely high due to the virtual environment's complexity and its impact on user experience and safety. Innovating in this space is a multidimensional puzzle. Early iterations of VR headsets, for instance, faced user discomfort and limited content, but these shortcomings led to significant advancements, paving the way for successful products like the Oculus Quest.

Setbacks in the metaverse are indispensable learning experiences, guiding developers through user interface design, realistic graphics, and intuitive experiences. Resilience and adaptability are vital for pioneers in metaverse innovation. Each "failure" is an essential waypoint on the road to captivating digital worlds.

Hackathons are an excellent means to foster safe spaces for innovation without fear of failure.[20] They mix fun and pressure, creating an environment focused on technology, problem-solving, and innovation. Netscape cofounder Marc Andreessen describes hackathons as "a science fair on steroids," a chance for people to come together, build something cool, and learn new things.

During a hackathon, a company may shut down for a week or more, focusing solely on innovating and presenting. Hackathons convene teams of developers, designers, and other professionals, encouraging interaction across different areas. This setting fosters inclusive, cross-group innovation, highlights new talent, and boosts team spirit and morale. Hackathons also enhance the company's brand, promoting an image of innovation and inclusivity.

The Failure Option

Resistance is no longer futile, and failure is an option. Leaders must effectively communicate that failure is an integral part of the innovation process, especially in the metaverse. The cutting-edge technology we enjoy today is the result of numerous failures. Where success is in an organization's DNA, so is failure. Embracing failure fosters the freedom to innovate, create, and invent, ultimately leading to market success.

Consider the tale of the Nintendo Virtual Boy.[21] Launched as a venture into virtual reality gaming, it failed due to its lack of immersion, poor responsiveness, and high cost. Its confusing operating system and limited game selection contributed to its downfall. This misfire could have been avoided if Nintendo had allowed more time for development and addressed critical feedback. The Virtual Boy serves as a valuable lesson for metaverse developers: prioritize readiness and user well-being to turn innovative ideas into successful realities.

Making it safe to fail is essential for businesses aiming for success in the metaverse. For instance, a software company developing a new social VR platform can quickly test and iterate on features without fear of breaking the platform. This approach ensures a platform that meets user needs. Similarly, businesses creating educational VR experiences can test and refine content without worrying about boring or confusing users, resulting in more effective and engaging learning experiences.

Winning businesses and products often use failure to their advantage. The journey from problem to solution to invention to product typically involves multiple failures. Underrepresented inventors may self-exclude due to misconceptions about patents and inventors. Additionally, inventors who believe in the myth of the eureka effect may freeze when everything doesn't come together instantly. Innovation is refined through failure, and since most valuable tech inventions are collaborative efforts, negative sentiments about failure can spread within teams. Misconceptions about the innovation process and fears of failure deprive the world of great products and services.

Messages from leadership can create effective narratives about inventing and innovating. The key message should be to try, to contribute ideas, and to understand that there are no guarantees. Innovative employees reflect the business in microcosm, with its risks and rewards. Embracing failure as a step toward success is crucial.[22] Jensen Huang, CEO of Nvidia, often reminds his employees that sitting on the innovation sidelines is not an option. Such messages and their inherent permissions must come from leadership, redefining the tone and extending new possibilities across the workforce.

Failure is not just an option; it is a welcome indication of movement toward success. Leaders must foster an environment where failure is seen as a natural part of the innovation process, enabling employees to explore, experiment, and ultimately succeed in creating groundbreaking products and services for the metaverse.

Not Failing, Not Trying

If you are not taking risks and experimenting, you are not likely failing, but you are also likely not innovating. This assertion is especially true when it comes to creating technology, content, and services for VR and AR in the metaverse. Companies need to be willing to try new things, even those they think will not work. New products, inventions, and breakthroughs in the VR and AR space often grow out of multiple efforts and repeated failures.

Take the development journey of the Oculus Rift, for example.[23] As one of the first commercially successful VR headsets, it stands testament to the importance of iteration and failure. Its founder, Palmer Luckey, went through the painstaking process of creating and discarding many prototypes, learning from each one that failed, before finally arriving at a version that was ready for the consumer market.

Similarly, consider the evolution of *Pokémon GO*.[24] The 2016 mobile AR game encountered numerous teething problems, such as server issues and bugs, yet it pushed through to become a landmark in AR gaming. Its more than 1 billion downloads in the first year did not come easy, but the lessons from its initial struggles helped refine the experience and maintain its popularity.

Lastly, look at Google Glass.[25] Although it didn't achieve commercial success, its journey wasn't in vain. It laid the groundwork for later AR devices such as the Magic Leap One and the Microsoft HoloLens. These successors drew invaluable insights from the shortcomings of Google Glass, leveraging its failures to guide their designs toward market acceptability and usability.

Failure is the biggest indication that you are trying and making progress. If you're not failing, are you really trying? Safe-to-fail is the ultimate green light. If no one is failing, especially at large organizations focused on innovation, then possibly not enough people are really trying. However, it must be the right kind of failure. Companies must foster an environment where failure is seen as a valuable part of the innovation process, allowing them to learn, adapt, and ultimately succeed in creating groundbreaking products and services for the metaverse.

Encouraging Intelligent Failure

What do we mean by "the right kind of failure"? Amy Edmondson's exploration of psychological safety in her latest book, *Right Kind of Wrong: The Science of Failing Well*, categorizes failure into three types: simple failure, complex failure, and intelligent failure.[26] Understanding these can help companies navigate the unique challenges of VR and AR development.

Simple failures in the metaverse context could be likened to basic software bugs or design oversights. While preventable, these failures provide immediate lessons. For instance, a VR application might initially cause user discomfort due to poor interface design. With proper feedback mechanisms, this can be quickly identified and rectified.

Complex failures arise from the intricate interplay of many subsystems within the metaverse. This could include interoperability issues between different VR and AR platforms, where a cascade of minor incompatibilities leads to significant reductions in user engagement or system functionality. As the metaverse grows, identifying and learning from these interdependencies will be crucial for developing robust and resilient systems.

Intelligent failures are the "good failures" that lead to significant breakthroughs. These involve trial and error in uncharted territories. An example in the metaverse is the initial public rejection of advanced VR headsets due to high costs and limited content. These setbacks informed manufacturers about consumer expectations, driving them to create more accessible and content-rich experiences.

Organizations that embrace failure, particularly intelligent failure, gain a competitive advantage. They encourage inclusive voices and remove barriers that prevent women and other underrepresented individuals from engaging in innovative thinking. These groups often fear that failure will confirm stereotypes about their identity or background, leading them to move cautiously in presenting new ideas. This cautious approach can stifle innovation.

To foster innovation, organizations need to create an environment where failure is not only accepted but encouraged. This involves recognizing that failure is a reliable indicator of effort and a necessary step toward breakthroughs. By allowing failure to flourish, companies can encourage diverse voices and innovative thinking, ultimately leading to more successful and inclusive outcomes in the metaverse.

Supporting Failure

As the saying goes, "Failure is success in progress." Though often attributed to Albert Einstein, its exact origin remains unclear. Nevertheless, the sentiment holds true—learning from failure is an essential part of innovation and growth. Creating a culture where failure is seen as an opportunity to learn and grow accelerates innovation and enhances employee creativity.[27] Embracing failure is essential for fostering a growth mindset.[28] When employees feel safe to fail, they can make unintended breakthroughs, as evidenced by the accidental inventions of penicillin, Post-it notes, Coca-Cola, Velcro, the microwave oven, and the World Wide Web.

A workplace culture that encourages failure is a counterintuitive but effective business strategy. Many employers recognize the power of failure in theoretically, but putting it into practice requires more than just endorsing the idea. It demands a complete cultural shift where employees truly feel safe to fail. Leaders must consistently communicate that failure is acceptable and not punishable.

Building something unprecedented, like the metaverse, necessitates accepting failure as inevitable. Poor communication about failure can drive innovators away, whereas celebrating "successful failures"—ideas that don't market well but hold promise—can signal progress. Examples like the Segway, early electric cars, and Betamax show how near misses can contribute to future successes. Patterns from failures can guide future processes, making them valuable learning experiences.

Failure should be acknowledged positively as a sign of effort.[29] Organizations can institutionalize a safe-to-fail culture by celebrating failures. Open communication from leadership and celebrating unsuccessful efforts foster psychological safety. Successful entrepreneurs like Walt Disney, Steve Jobs, Jeff Bezos, and J. K. Rowling faced multiple failures before achieving success, illustrating the importance of resilience.

Without support for failure, decision avoidance or paralysis can set in. Fear of negative repercussions like lack of promotion, reduced respect, or termination can stifle innovation. Companies must cultivate an environment where taking chances on potential failures is accepted and encouraged, ensuring that employees do not equate failure with personal inadequacy.

Notable Failures

The journey of innovation is often marked by failures that pave the way for future successes. The Newton personal digital assistant (PDA) was an early example, struggling with poor handwriting recognition, short battery life, and high cost, leading to its commercial failure.[30] Similarly, HD DVD, introduced in 2006, lost the format war to Blu-ray due to limited studio support and higher costs.[31] The Windows Phone operating system, despite its potential, failed between 2010 and 2017 due to low app support, an unfriendly user interface, and poor marketing.[32]

The AR and VR sectors are littered with similar stories of failure and lessons learned. Google Cardboard, launched in 2014, aimed to make VR accessible with a low-cost headset, but it was uncomfortable and offered a limited field of view.[33] The Oculus Go, a standalone VR headset, struggled with a limited content library, leading to its demise.[34] Google Daydream, introduced in 2016, also suffered from limited content and poor marketing.[35] The HTC Vive faced challenges with its high cost and complex setup, while Microsoft HoloLens was hindered by its price and limited field of view.[36]

These failures are part of a larger pattern in the tech industry, particularly in gaming, where trial and error are common. Nintendo's aforementioned Virtual Boy, for example, was a notable flop in the mid-'90s. It was released into a saturated market dominated by SNES and Sega Genesis, and its high price, uncomfortable design, and low-resolution graphics led to its quick failure. Despite this, the Virtual Boy's technology laid the groundwork for future VR headsets like the Oculus Rift and PlayStation VR.

Attempts to deliver movies via VR headsets have also seen limited success. Products like I-Glasses (no connection to Apple), Sega VR, DAQRI, and Magic Leap's headsets faced high costs and technical limitations. Microsoft's HoloLens and Google Glass also fell short in the market due to similar issues.

The I-Glasses, released in 2008 by I-O Data Device, were pioneering but hampered by high costs and poor graphics.[37] Sega VR, introduced in 1991 for arcades, failed due to its high price and a limited game library.[38] Daqri's AR headsets and Magic Leap's products suffered from technical problems and unmet user expectations, causing issues like motion sickness and eye strain.[39]

Despite these setbacks, each failure has provided valuable feedback for the industry, helping to guide future innovations in AR and VR. These experiences highlight the importance of iteration and resilience in technological development, demonstrating that even significant failures can drive future success and advancement in the metaverse and beyond.

Intuitive Thinking

A successful product in the VR market is often about a decade in the making. While it has only recently begun to gain traction, the metaverse concept has been around for many years. Similarly, AI, AR, VR, and blockchain, although recently commercialized, have been subjects of research and development for decades. These technologies have all faced numerous hard-won successes and disappointing failures along the way. The path forward has often been unclear, requiring innovators to rely on intuition to identify new problems, opportunities, and solutions.

Intuition has played a critical role for many creative individuals and inventors, helping them break free from traditional thinking patterns and arrive at innovative solutions. Leaders of companies building the metaverse need to foster a culture that values and encourages intuitive thinking, integrating and leveraging the right kinds of failure.

According to Daniel Kahneman's theory of fast and slow thinking, for which he won the 2002 Nobel Prize in Economic Sciences, intuitive thinking is fast but prone to error, while rational, slow, deliberative thinking is typically more valued in business.[40] However, in the context of innovation, particularly in emerging fields like the metaverse, intuition can be invaluable. Organizational culture must adapt to support intuitive thinking alongside rational analysis.

For decades, the metaverse, AI, AR, VR, and blockchain have been fields of trial and error. Despite numerous blunders, these technologies have evolved, often driven by intuitive insights. Innovators and leaders must recognize the importance of intuition in navigating the complexities of these technologies. By valuing both intuitive and rational thinking, companies can create environments where creative processes flourish and lead to groundbreaking innovations.

Ensuring Psychological Safety

Any innovation, it should be noted, comes from a place of creativity. Successful collaboration relies on psychological safety.[41] Without it, efforts to foster creativity or support experimentation may fail from the start. Employees need to feel comfortable sharing their ideas, regardless of how unconventional they may seem. Psychological safety at Google, for instance, has been credited with the development of Gmail and Google Search.[42] Research shows that Google teams with high levels of psychological safety are among the most productive and innovative. At Pixar Animation Studios, psychological safety encourages employees to share their ideas, even if they are not fully formed. This culture of open feedback has led to beloved animated films like *Toy Story* and *Finding Nemo*. IDEO, a design and innovation firm, considers psychological safety a core value, believing that everyone has the potential to be creative.[43] This environment has contributed to the development of products like the Nest Thermostat and Fitbit fitness trackers.

Making it safe to fail is crucial for injecting the psychological safety needed to encourage experimentation.[44] Many metaverse technologies will go through multiple generations and iterations, marked by a continuous stream of failures, successes, and improvements. Punishing failure can stifle innovation, whereas making it safe to fail requires businesses to trust their workforce. Leaders must be comfortable and confident that employees are genuinely focused and trying their best, not using the safe space to reduce effort. In this sense, leadership may be in of need as much psychological safety as staff. This mutual trust fosters an environment where innovation can thrive, allowing companies to navigate the challenges of developing new technologies for the metaverse.

The Role of Leaders

Creating a culture where employees feel safe to take risks, experiment, and learn from their mistakes is a crucial responsibility for leaders. They can start or sustain this process by admitting their own mistakes, which fosters an atmosphere of openness and mutual respect.[45] Leaders who model the behaviors they want to see and establish systems and processes that support risk-taking and innovation help instill psychological safety within their organizations. This cultural shift must be backed by providing the necessary resources, training, mentorship, and financial support that innovators need to succeed and turn the philosophy of safety into a daily reality. Leaders can also provide support by being available to answer questions and offer guidance. Celebrating successes and failures further shows employees that it is okay to fail. Celebrating mistakes provides much of the attention required to learn and grow.

Satya Nadella has fostered a "growth mindset" culture at Microsoft by supporting employees in risk-taking and experimentation, even if they fail.[46] He ensures that employees have the resources and support they need to succeed. Microsoft's Fail Fast program provides funding and support to allow employees to experiment with new ideas. Jeff Bezos also supports experimentation and failure at Amazon, believing that "failure is an invention and discovery process." [47] Amazon's Challenge Teams program encourages employees to work on innovative projects. Elon Musk of Tesla and SpaceX insists that "failure is an option here. If things are not failing, you are not innovating enough."[48] Tesla's Test Lab enables employees to test new ideas in a controlled, safe environment.

Leaders must make it safe to fail to enable businesses to fearlessly innovate. Achieving such an environment requires removing the fear associated with performance, praise, consequences, and job security. Persistent messaging from the top, such as daily video emails about ongoing initiatives, whether successful or not, can help establish and sustain a culture where innovators feel empowered to take new steps and achieve unexpected results.

Right Message, Right Action, Line Up All Innovators

Communicating a message to staff to value and share their ideas is key to broad engagement and innovation.[49] Leaders must go beyond sending memos if they want to sustain connection; they need to "come down" and engage directly with their employees, embodying and modeling the changes they want to see.

Leadership should set the tone and issue the invitation, understanding that employees may be skeptical of requests for engagement. They may be accustomed to requests for all-hands engagement and may be reluctant to interpret the message from leadership as more than the standard cliché. Thus, the message from leadership must resonate with authenticity and vulnerability.

Getting things right in the metaverse may involve numerous failures. Leaders must make it safe to fail and keep trying, sharing their own experiences as innovators. Whether they have patented innovations or offered creative strategies, leaders are innovators. Sharing their stories of attempts and failures empowers their messages. When senior leaders tell their stories, the potential to engage others increases exponentially. Turning their "I," "me," and "my" into "we," "us," and "our" helps upper leadership connect with employees and move the needle on engagement.[50]

Leaders must remind employees why now, more than ever, everyone needs to contribute. The metaverse must do better than the internet has done for nearly three decades, building without the barriers and biases that bogged down Web 1.0 and 2.0. Businesses can create inclusive metaverse experiences by fostering an inclusive mindset among designers, engineers, marketers, and other employees. Communication from the top can remind everyone that their ideas are valuable and welcomed, demonstrating the regard and inclusion the business needs to emulate.[51]

Throughout this messaging, remember that psychological safety is crucial for creativity and innovation. Employees need to feel safe to fail, particularly underrepresented innovators who may have different sentiments about failure.[52] Building the metaverse—software, hardware, networks, and content—without inclusive input and innovation risks failure and loss of investment.

The media and settings used by leadership to transmit their messages may vary. One-time meetings can serve as launch points but well-crafted short videos allow important messages to be stored, distributed, and accessed multiple times. Internal communications teams should rework and deliver messages from top executives without undermining authenticity. Leaders should stay engaged with the messaging to avoid losing the opportunity for personal connection.

Any indication that the leader's ideas and underlying sentiments have been standardized, reworked, or repackaged undermines authenticity and nullifies the leader's intentions. As witnessed at a large technology company, internal communications teams may make the mistake of characterizing the push for inclusive innovation as a need to reshape behaviors toward marginalized groups. When this happens, fatigue replaces enthusiasm. Campaigns fail. Connection is lost and may be difficult to reestablish.

When the goal is innovation—*breakthrough* innovation—the most critical cache of data and information relates to your own workforce. These are the innovators, the inventors, the problem-solvers. How are they doing? Leaders need to take the baseline state of their employees in order to communicate with them effectively and motivate their innovation efforts.

As momentum toward the emergence of the metaverse continues to build, communication and encouragement will be key. Organizations may bounce between inspiration and anxiety, and the latter is a known enemy of innovation. Leadership communication is critical to keeping people motivated and focused. Leadership messaging can sustain the workforce through the need to innovate and commercialize rapidly, keeping them motivated and focused amid the innovative demands of the metaverse.

6 Designing for Impact

There is a path from wanting an inclusive metaverse to building one. Leaders focused on creating the metaverse must design, communicate, and implement an inclusive innovation strategy.

Building a dynamic and future-ready metaverse requires more than intent—it demands a structured strategy that translates vision into action. Leaders shaping the metaverse must design, communicate, and implement an innovation strategy that ensures broad participation and engagement. Success in the metaverse depends on fostering an environment where all contributors feel empowered to share ideas, challenge assumptions, and collaborate effectively.[1] A thriving innovation culture is one where employees feel valued and respected, regardless of background, abilities, or perspectives.[2] Without this, businesses risk missing out on valuable insights, limiting creativity, and ultimately weakening their competitive edge.

The shape of a business and its innovation reflects its leadership's communication. Engaging the workforce in the innovation necessary to compete in the metaverse is akin to throwing a party. If no one shows, you face embarrassment and loss. If a business fails to engage its workforce in proper innovation, the impact can be significant. Large businesses may shake off the loss, but smaller organizations can be seriously rattled or perish. Leaders must communicate effectively to bridge the gap between them and their workforce to ensure employee engagement in the innovation process.

Decide Where to Play and Innovate

Assessing core competencies and choosing a focused niche allows businesses to become experts in their field and develop a competitive advantage. For example, Apple has concentrated on a limited number of products—the iPhone, iPad, and Mac—building expertise and a loyal customer base. This focus enabled resilience when Apple scrapped the "Apple Car" project in early 2024 after years of investment.[3] Amazon's core focus on e-commerce, Microsoft's on productivity software and cloud computing, Google's on search and advertising, and Tesla's on electric vehicles illustrate how businesses succeed by sticking to their strengths.

Businesses preparing to enter the metaverse should similarly focus on areas just beyond their core competencies. By analyzing their current customer base and identifying new ways for those customers to use existing products and services, businesses can find viable strategies in the metaverse. These strategies can involve real-life services such as banking, real estate, and counseling, or virtual equivalents of real-life products and services. Essentially, what a company does well in real life can translate into the metaverse with minimal reinvention or upskilling.

For instance, in 2022, JPMorgan Chase launched a "lounge" in Decentraland, showcasing its blockchain achievements and presenting videos on e-commerce and financial technology. JPMorgan Chase envisions offering traditional banking services like credit, mortgages, and rental agreements in the virtual world, merging physical and virtual financial realms.[4]

Adapting and innovating are crucial for businesses to stay ahead in the rapidly changing metaverse landscape. Competition will be fierce as companies from around the world will compete for attention and market share. Whether a massive tech company or a small family business, each must craft its unique approach to succeed in the metaverse.[5] This requires innovative thinking and problem-solving. Regardless of size, opportunity, or resources, any business can formulate, execute, monitor, and tweak an effective innovation plan to thrive in the metaverse.

Decide What Success Looks Like

Success varies by industry, goals, and development stage. For a small company entering the metaverse, success may mean reaching a certain number of users or generating specific revenue. Mature tech or retail companies may define success as maintaining market leadership or developing new innovative products. Measuring achievements against an established strategy brings clarity and focus. Many organizations lack clear success metrics, making it hard to gauge progress. Positive earnings alone may not indicate

Designing for Impact

success. In the metaverse, especially early on, businesses need well-defined indicators of success to navigate and measure progress.

Health care is an example of defining success factors. In the health care metaverse, 81 percent of leaders expect positive organizational changes, including reducing gender and racial gaps in patient treatment.[6] This digital space offers new ways to train health care workers, such as practicing surgery virtually, and makes learning more accessible to remote students. Currently, over 90 percent of health care centers offer remote consultations, and the metaverse will enhance these opportunities.[7] New technologies will allow health care professionals to explore 3D models of patient bodies, improving diagnosis and treatment. Success in health care includes promoting gender equality and reducing inequalities. The metaverse can help female health care workers, who face social or cultural restrictions, access equal work and learning opportunities. It can also offer specialized health care services virtually, reducing the gap between urban and rural health care access.

Surgeons can now receive immediate support from global experts during complex operations. Wearing VR (virtual reality) headsets, they can transmit their field of view to remote specialists, who can then provide guidance. This advancement improves surgical accuracy and patient safety, especially for rare or complex procedures requiring specialized knowledge not available locally.[8]

The partnership between Apollo Hospitals and American Tower Corporation (ATC) in India exemplifies innovative approaches to addressing health care disparities in rural areas through technology. By launching five digital pharmaceutical dispensaries in rural regions, they are bridging the gap in health care accessibility for isolated communities. These dispensaries offer primary, preventive, and specialty teleconsultation services to around 250,000 residents, combining Apollo Hospitals' medical expertise with ATC's technological infrastructure.[9] This initiative allows remote communities to access quality health care without long-distance travel, showcasing how a successful venture between the virtual and physical worlds can create a metaverse that serves everyone.

Multilayered, Flexible Strategies

To navigate the competitive landscape of the metaverse, a multilayered and flexible strategy is essential. Companies must align their long-term vision with what they aim to achieve within the metaverse, setting specific, near-term goals and objectives complete with a timeline for realization. Adapting to inevitable and unpredictable changes requires a strategy that is not static but one that undergoes regular reassessment. A large business offering multiple products or services needs several innovation strategies upon entering the competitive landscape of the metaverse. This is akin to professional sports teams having different playbooks and game plans that vary by venue or competition, all featuring significant flexibility. Winners in any field prepare for changing conditions, and in the interconnected ecosystems of the metaverse, disruptions are inevitable. A network company may lag in innovation or essential partners may leave, so sustainable strategies must account for these unexpected changes.

Microsoft's introduction of Mesh for Microsoft Teams in 2021 is an example of how traditional business tools are adapting to the metaverse. By integrating Mesh with its standard products and services, Microsoft facilitated the future of work by enabling virtual collaboration, training, expertise sharing, and 3D-space designing. This move illustrates the necessity of blending current offerings with new technologies to remain competitive.

Accenture's adoption of One Accenture Park for onboarding new employees illustrates how virtual platforms can be used to integrate new hires, with over 300,000 individuals onboarded in this engaging virtual space.[10] This process transcends traditional methods by immersing recruits in a virtual duplicate of Accenture's physical offices, adding layers that will potentially be part of the future metaverse features. This represents a distinctive use of the metaverse for corporate processes, different from the norm of online onboarding.[11]

Takeda, the pharmaceutical company, is another example of how Mesh can be used to illustrate global connection and cultural development within a corporation.[12] The Hirameki Garden is a virtual space that mirrors Takeda's values, hosting large events and reflecting the company's commitment to fostering a strong corporate culture and driving innovation. Leo Barella, Takeda's chief technology officer, emphasized the burgeoning possibilities of Mesh, pointing out that they are only beginning to tap into its vast potential for custom virtual experiences.[13]

BP has partnered with Microsoft to create Highly Immersive Visualization Environments (HIVEs), which exemplify effective and eco-friendly ways teams can collaborate globally.[14] The HIVEs project, envisioning virtual

Designing for Impact 99

gatherings for monitoring remote wind turbines or troubleshooting equipment repairs, underlines the transformative capabilities of the metaverse for global energy operations. This technology is key to accelerating informed and safer decision-making in alignment with BP's integrated energy ambitions.

Lastly, the Mercy Ships case exemplifies humanitarian and educational applications of metaverse technologies. Mercy Ships is a nonprofit organization that operates the world's largest civilian hospital ships, providing free surgical care and medical training to communities in low-income countries, primarily in Africa. To enhance operational effectiveness and improve volunteer preparedness, the organization has introduced the *Global Mercy* digital twin—a virtual replica of its newest hospital ship, *Global Mercy*. The *Global Mercy* digital twin allows for more efficient training of volunteers and enhances operational effectiveness.[15] This virtual experience ensures that volunteers are well prepared before they even step foot on the actual ships, signifying a shift in how charitable organizations can utilize the metaverse.

These examples demonstrate the diverse ways companies are leveraging the metaverse to innovate and stay competitive. By embracing flexible and inclusive strategies, businesses can navigate the complexities of the metaverse and drive success in this evolving digital frontier.

The Value of IP

Owning, controlling, and uniquely identifying innovation is a crucial economic engine. Users may not consider patents or trademarks when using an iPhone or Xbox, or sipping a Coke, but businesses are always aware of their intellectual property (IP). Inclusive innovation often leads to patent filings, and IP will be vital in the metaverse. Regardless of their role, all businesses in the metaverse are likely to create intellectual property. Ensuring employee engagement at all levels fosters inclusive creative outcomes.

However, the value of IP is often misunderstood by departments not directly involved in protecting it, leading to tensions within organizations. I recall a time when I was working with a company that had just developed a groundbreaking technology. The engineering team was thrilled with their innovation, but when it came time to patent the invention, they were met with resistance from the marketing department. The marketing team didn't fully grasp the significance of securing IP rights and saw the patent process as a hindrance to their aggressive go-to-market strategy.

The engineers were frustrated, feeling that their hard work was being undervalued, while the marketing team felt bogged down by what they perceived as unnecessary bureaucracy. It took several meetings to bridge the gap, emphasizing the critical importance of protecting innovation—not just for immediate market advantage but for the long-term value it would add to the company. Eventually, a strategy was agreed upon that allowed the marketing team to move forward with their plans while ensuring the IP was properly secured. This experience highlighted the importance of cross-departmental communication and education about the value of IP, ensuring that all teams understand how their work contributes to the company's broader goals.

The Subtle Art of Communication

Let's look at an intriguing challenge my team and I encountered during a consulting engagement at a large global tech company. The company was eager to broaden its innovation pipeline, particularly by drawing in more women and minorities. However, despite its best efforts, something wasn't clicking. Even when underrepresented individuals entered the innovation process, they weren't transitioning into patented inventors. The numbers told a stark story: across its global workforce, only 15 percent of patent-earning innovators were women, even though women made up between 22 to 37 percent of the company's full-time tech staff.

As we delved into the situation, it became clear that the issue wasn't a lack of talent or ambition, but rather something subtler—something that lay in the company's communication methods. While leadership aimed to make the patenting system accessible to all, data showed variations in engagement. Male employees were more likely to visit the company's patent website and navigate the process independently, whereas women and employees from underrepresented backgrounds often engaged after speaking with a member of the patent staff.

It wasn't just about access; it was about how the information was being communicated and received. The company's patent culture and communication styles were inadvertently geared toward certain social and communication norms, which didn't necessarily align with how different employees processed and acted on information. The result? A significant portion of its talent pool wasn't fully engaging with the innovation process.

Recognizing this, we knew we had to reframe the strategy. The goal wasn't simply getting more diverse employees involved—it was about developing a communication approach that would engage the entire organization. Instead of targeting messages specifically by demographic groups, we needed to cultivate a shared purpose that inspired participation across all employees.

To make this happen, we had to convince the leadership communication team to rethink their approach. I remember the long nights we spent drafting a memo—a hefty 50-page document—detailing why a broader, engagement-driven communication strategy was essential. We needed to explain why targeting messages specifically to race and gender could be counterproductive, and how success would only be achieved by bringing everyone, regardless of their background, into the shared purpose of the organization.

This was no easy task. It required a subtle art, a delicate balancing act. I had to make the leadership see that while their intentions were good, their methods were unintentionally creating silos rather than breaking them down. The memo was filled with data, yes, but it was also filled with stories—real examples of how other organizations had successfully shifted their approach to communication in a similar way and saw tangible improvements in innovation.

In the end, we didn't just tweak the message; we overhauled it. I worked closely with the company to change the structure, content, and flow of information. We made sure that everyone, especially those from historically excluded groups, could see themselves as potential innovators. This wasn't about isolating groups with targeted messaging but about crafting a message that resonated with everyone.

We developed a communication strategy that included direct input from leadership and showcased diverse inventors from across the company in short videos. These videos featured people from various roles, backgrounds, and career levels sharing their journeys through the patent process. The idea was to break down the narrow stereotype of what an innovator looks like and ensure that employees saw innovation as a collective opportunity rather than an exclusive process.

But even with all these changes, I knew there was more to be done. Communication, as well intentioned as it may be, can sometimes miss the mark in understanding the subtle nuances of transformational systemic change. There's an art to engaging everyone—not just by targeting specific groups but by fostering a collective sense of purpose that transcends individual differences.

That's why I took the time to write that 50-page memo and have multiple conversations with various communication team members detailing why this approach was necessary. The goal wasn't just to make a case for change—it was to guide leadership toward seeing the bigger picture. I had to show them that success wouldn't come from narrowly segmenting outreach by demographic groups but rather from embedding a culture of innovation that resonated with everyone and aligned with the purpose of the organization. Internal communication teams, despite their best intentions, using their resources and employed mechanisms can sometimes oversimplify issues and lead to misapplied solutions. True success isn't about increasing diverse participation at the expense of others; it's about creating a balance where everyone is engaged and valued.

Ultimately, it's about creating a culture where everyone feels they belong and can participate in the innovation, where every voice is heard, and where innovation can truly thrive. Success, I explained, looks like everyone being at the table, with no one relegated to the margins. It's a subtle but crucial shift in thinking—one that, when embraced, can lead to real, lasting change.

The leadership embraced this new approach, and slowly but surely, we began to see the results. More people were engaging in the innovation process, patents were being filed by a more diverse group of employees, and the company saw firsthand the value of a broader, engagement-driven innovation strategy. It was a challenging but rewarding journey, proving that reframing communication can unlock hidden potential, expand innovation, and drive real business success.

Assessing Engagement

Engaged employees are more likely to be creative, innovative, and committed to the company's success.[16] They are also more likely to strive to make the metaverse a welcoming space for all consumers and an economic success for their company. Assessing employee engagement is crucial for identifying areas where the company can improve its innovation process.[17] Employees who are not engaged may feel their ideas are undervalued or that they lack the resources to develop and implement them.

Understanding employee sentiment provides a baseline to identify and address potential biases in the company's invention and innovation processes. However, getting an entire workforce to think of themselves as innovators is challenging without first understanding how they feel about themselves and their past experiences with the company's innovation processes. Leaders should gather data to determine how engaged staff are in these processes. Analyzing this data offers insights into what has worked and where roadblocks exist, making it an essential step in reconfiguring the perceptions and behaviors of both willing and reluctant innovators.

I recall a specific instance at a large tech company where the leadership was making significant moves, laying the groundwork for a major strategic shift. During and after an all-hands meeting, the frustration and anger among employees was palpable. They didn't understand the direction the company was taking, and the communication team seemed oblivious, congratulating themselves on a job well done. This disconnect was a clear indicator that the communication strategy was not resonating with the workforce.

These experiences underscore the importance of cross-departmental communication and the need for leadership to be attuned to the concerns and perceptions of their employees. By listening closely and adjusting its messaging, the company was able to realign its communication strategy, ultimately fostering a broader, engagement-driven environment for innovation. In the end, success in engaging a diverse workforce is not just about crafting messages, it's about ensuring those messages resonate, inspire, and align with the company's broader goals.

Is Everyone Here . . .

Inclusive design and innovation isn't just about policy—it's crucial for a business's revenue and product viability. Developing technologies and services that resonate with a broad customer base ensures greater adoption, market relevance, and long-term viability. Companies that engage a wider range of innovators, including those from traditionally overlooked roles such as customer support, often uncover valuable insights that lead to product improvements and expanded market appeal.

The task of developing a metaverse that reflects and serves a broad range of users comes with both challenges and opportunities. Currently, leadership roles in this space are largely concentrated within a specific demographic, which can naturally influence decision-making perspectives. To stay competitive and drive innovation forward, businesses can benefit from expanding engagement beyond traditional talent pools, not just through structured initiatives but by reassessing how innovation is encouraged and supported across teams. Differences in patent filings and innovation contributions among various groups suggest that there are untapped opportunities for market growth, business expansion, and technological advancement. A wide range of perspectives and experiences is essential to developing a flexible, future-ready metaverse. Organizations that broaden participation and create space for more contributors can gain a competitive edge, drive fresh thinking, and strengthen their innovation pipeline. As digital ecosystems evolve, businesses that embrace new ways of engaging talent at every level will be better positioned for long-term success.

Creating an environment where employees feel encouraged to share ideas and contribute leads to more scalable, competitive, and widely adopted solutions. Expanding participation isn't just about who is involved—it's about ensuring companies harness the full creative and problem-solving potential within their workforce. This is why allyship is essential. Those in leadership and influential positions have a unique opportunity to amplify ideas, remove barriers, and champion a culture of engagement that benefits the entire organization. When leaders actively support broad participation and collaboration, they create an environment where innovation thrives and businesses unlock their full potential. True progress happens when everyone feels empowered to contribute, and those with influence use their position to open doors for others.

. . . Really Here

Reducing feelings of exclusion in the innovation process can lead to unexpected benefits.[18] Regularly assessing employee engagement and perceptions helps organizations make necessary adjustments to harness their full innovative potential. Diverse individuals bring expertise and ideas that have been tested and vetted, adding value to the innovation process.

The metaverse can empower individuals with cognitive or physical challenges, enabling them to rejoin the workforce in meaningful ways, such as remote inspections through digital twin technology.[19] Examples like Floreo,

Designing for Impact

which uses VR to help neurodivergent individuals navigate daily life, and Vision Buddy, which creates VR headsets for senior citizens to enhance independence, show how thoughtful innovation can have a broader audience while unlocking new market potential.[20]

Organizations that close participation gaps—whether in talent pipelines, product design, or innovation culture—position themselves for long-term success. This approach is not just about expanding access but about ensuring that products and services resonate with a wide market. A well-developed innovation ecosystem lays the foundation for a metaverse that reflects its users, requiring both intentional strategy and effective communication to bring those ideas to life. In the globalized, interconnected landscape of the metaverse, companies with diverse teams are better positioned for intellectual property development, commercialization, and market leadership. By ensuring that all contributors are fully engaged in the innovation process, businesses gain a competitive advantage that drives both creativity and commercial success.

Talking Innovation

Consistent communication about innovation is crucial for breaking down stereotypes and default thinking.[21] Leaders should remind employees that everyone involved in creating the metaverse is working for future generations. They should emphasize that while getting it right is the goal, perfection isn't expected, especially with early product introductions. The modern product rollout allows for extensive feedback from users, meaning that engineers, designers, marketers, and decision-makers are as integral to the process as the users who offer praise or criticism.

Employee engagement in innovation is directly tied to how well they understand the organization's goals. Since knowledge levels vary across a company, establishing information hubs is essential.[22] These hubs should provide all employees with easy access to the information they need to see their role in innovation. Many employees might view innovation as something others do, seeing the process as exclusive or opaque. They need guidance on the legal aspects of inventing and patenting and advice on how to collaborate effectively within the organization.

Information hubs should go beyond just collecting ideas; they should help employees redefine their identities within the innovation process and offer detailed, practical guidance. While patenting remains a significant milestone, innovation extends far beyond patents, encompassing problem-solving, process improvements, and creative advancements. Effective messaging and the right messengers are key to transforming perceptions and cultivating a thriving culture where contributions at all levels drive progress and success.

You Got the Power

Organizations can unlock their full innovative potential by analyzing data and feedback to make small but impactful adjustments. When employees feel valued and view themselves as part of the company's collective innovation resource, a shift occurs. More employees will begin to see themselves as inventors and share ideas if the company makes its innovation and patent processes accessible and easy to understand. Encouraging this mindset is half the battle; once achieved, the organization should highlight the stories of peers who have successfully navigated these processes.

To sustain this momentum, internal processes must be redesigned to meet all innovators where they are, ensuring everyone has equal access to information about how innovation happens and where it can lead. Innovation and problem-solving naturally lead to inventing, which may lead to patenting under the right conditions. Clear communication about these processes can improve workplace attitudes toward innovation and boost employee optimism.

For innovation to thrive, it must be perceived as fun and engaging. If the process feels too intimidating, exclusive, or bogged down in legal jargon, many potential innovators—especially women and minorities—may hesitate to participate. Psychological safety is key; employees need to feel that innovation is an enjoyable, accessible part of their work to fully engage.

In Your Hand

A company should create a centralized portal where employees can anonymously submit ideas, ensuring they are evaluated purely on merit. A transparent process for assessing these ideas, coupled with regular status updates and a consistent reward structure, will break down barriers and foster greater inclusion in innovation, inventing, and patenting. The potential of

Designing for Impact

the metaverse, supported by employee engagement data, helps businesses identify new opportunities and move forward with confidence.

Incorporating everyone into the innovation process starts with effective communication. Understanding the baseline state of innovation within the company allows leaders to craft messages that resonate authentically with a wide audience. Traditional communication methods may no longer suffice for a company aiming to lead in the metaverse. Leaders can only optimize their communication and fully engage their innovation resources by understanding the barriers employees face and their experiences with the company's IP processes.

Innovate from Within

Someone at your company holds the key to a breakthrough that could propel your business to industry leadership. But who? It could be anyone—even those in nontechnical roles. To stay competitive in the metaverse, businesses must innovate inclusively by engaging every employee.

For example, Nvidia Omniverse and artificial intelligence (AI) enabled BMW to streamline its factory planning process, leading to the launch of the world's first virtual factory.[23] This innovation allowed BMW to optimize its operations years before the physical plant in Debrecen, Hungary, opens in 2025, resulting in greater efficiency and sustainability.[24]

The metaverse requires extensive innovation across multiple sectors, from developing advanced headsets and haptic devices to creating faster, more reliable networks. While partners and outsourcing can help, businesses already have immense creative and problem-solving potential within their workforces. Leveraging this untapped talent is crucial for gaining a competitive edge in this rapidly evolving landscape.

Building Collaborative Efforts

The metaverse will be built on AR (augmented reality), VR, AI, and other technologies that are still in the early stages of their development. Bringing these technologies to maturity will require global collaboration among researchers and engineers. Additionally, creating engaging metaverse experiences for all users will demand the combined efforts of designers, writers, and artists to develop new forms of entertainment and education.

Collaboration isn't just important—it's essential, and it must be the default approach.

The importance of collaboration in driving successful innovation is so well recognized that many organizations are appointing "chief collaboration officers." These leaders often have dual roles: outwardly, they help businesses navigate crucial partnerships and ventures with vendors and other collaborators; inwardly, they foster intracompany connections. For example, at the venture fund Superset, this position is held by K. C. Maxwell, who plays a pivotal role in weaving together interests and expertise within the company. This role is valuable in both small organizations and large, decentralized businesses, which often need an internal system or referral agent to match talents and interests in ways that enable and accelerate innovation across the company.

Reaching Your Innovators

Promoting collaboration, valuing inclusion in innovation, recognizing and rewarding problem-solving, and providing essential resources are key strategies for businesses to engage all their innovators. Celebrating diversity in innovation means honoring the range of perspectives at the table. Companies can foster this by creating innovation groups, offering training, hosting challenges, setting budgets, and establishing councils dedicated to innovation.

The typical organization possesses a surprising range of expertise, often spread across various business units and sometimes in unexpected places.[25] Innovating and inventing are not exclusive to engineers and designers. For example, Apple's executive Ron Johnson came up with the idea for the Genius Bar, which significantly contributed to the iPhone's success.[26] Similarly, a 3M sales team played a role in developing the Post-it note, and customer service teams at Amazon frequently contribute valuable innovative ideas.[27] Tapping into all areas of expertise, including operations, finance, HR, and even outside vendors, offers businesses significant advantages.

Successfully innovating for the metaverse requires the diligence of teams with expertise that companies may overlook. A successful app or platform must be visually appealing, psychologically inviting, and backed by optimized technology. Business models must be smart and effective, and offerings must be compliant with regulations. By the time a product reaches

Designing for Impact

users, it has drawn on a broad spectrum of talent within and beyond the company where it was created. While partnering or outsourcing can supplement resources, companies that fully leverage the innovative potential of their workforce will hold a competitive edge in the metaverse.

The Need for Broad Engagement

Engagement in the metaverse, regardless of a business's specific focus, is a complex endeavor. Every action in this interconnected space can influence other parts, and missteps can create significant ripple effects. To successfully innovate and sustain the metaverse, businesses need to engage all eyes and hands.

Expanding participation naturally fosters further engagement.[28] For instance, someone with social anxiety disorder might use VR to safely practice social interactions, while someone else with chronic pain could turn to the metaverse for distraction and relaxation, reducing their dependence on medication. Individuals with mobility challenges can use virtual environments to practice walking, and those with dementia can navigate their neighborhoods virtually without the risk of getting lost. As users encounter different needs in the metaverse, they will drive innovation and policy development in response.

Different user groups bring unique concerns and priorities that shape innovation. Parents of young children may prioritize safety tools that prevent cyberbullying and inappropriate content exposure.[29] Some users may seek features that promote personal security and prevent harassment, while others may advocate for better customization of sensory experiences, such as noise control in virtual environments. These diverse perspectives generate new ideas, fueling continuous improvements across platforms and applications.

Since the metaverse is intended to serve a broad audience, involving a wide range of talents and perspectives in the innovation process ensures that apps, platforms, and policies are accessible, functional, and equitable for all users.[30] Many companies already possess the creative capital to achieve breakthroughs, but the challenge lies in engaging all voices. The language used to invite employees to participate in innovation is crucial. Word choice can influence how employees perceive innovation opportunities. For example, using a term like "sharing" might encourage broader participation than terms like "brainstorming" or "harvesting," which may carry unintended connotations. Effective communication can help make the challenges of innovation and problem-solving less daunting and more accessible.

The Right Messaging

Crafting the right messaging is not a one-size-fits-all endeavor. Messages must be carefully framed and adapted to resonate across diverse audiences, backgrounds, and cultural contexts. Even with the best intentions, communication can misfire, leading to unintended consequences.[31] For example, complimenting individuals as "articulate" or "the exception" may be intended as praise but can instead be perceived as a subtle dismissal of their expected competence. Similarly, phrases like "be more assertive," "take initiative," or "think outside the box" may not hold the same meaning across different cultures and work environments. When leadership messaging is misinterpreted, it can create hesitation, disengagement, or even defensive responses.

Strong leadership messaging is a foundation for expanding broad engagement, but simply stating that "everyone is an innovator" is not enough. The real challenge lies in overcoming the "Who? Me?" barrier—a mindset shaped by personal experiences and societal expectations that may lead some employees to self-exclude from the innovation process. Companies often reinforce these perceptions by unintentionally framing innovation as the domain of specific roles, backgrounds, or skill sets. To fully unlock innovation potential, businesses must actively dismantle these assumptions and guide employees toward redefining their role in shaping new ideas.

Leaders play a crucial role in setting the tone and inviting everyone to participate in innovation. Their messages resonate when they clearly articulate the challenges ahead, emphasize that innovation can come from anyone, and encourage persistent effort. By creating an environment of psychological safety, leaders can help individuals who don't see themselves as inventors feel comfortable sharing their ideas.

Take, for instance, the CEO of a major tech company who had long been a polarizing figure—both loved and hated by the media. Despite this, there was widespread admiration for his innovative strategies. Thanks to a new communication team, the CEO began to behave in a way that was authentic to his true self regarding innovation and how diverse it needs

Designing for Impact 111

to be, which, in turn, created more genuine engagement among employees. The ability to craft messages that resonated on a personal level was instrumental in turning around the perception of leadership and fostering a more engaged workforce. It also reshaped the company's internal culture and improved internal morale as well as its external reputation as an innovation-driven organization.

Purpose-Linked Campaign

Designing a campaign linked to purpose is essential for creating a cohesive and motivated workforce.[32] When employees understand how their contributions align with the company's overarching mission, they are more likely to engage deeply and innovate with intention.[33] A purpose-driven campaign connects individual roles to the broader goals of the organization, fostering a sense of belonging and shared responsibility. By clearly communicating the purpose behind initiatives, leaders can inspire action, drive collaboration, and ensure that every team member is aligned with the company's vision for success.

Demystifying Patents

Innovators in large companies who are women and minorities often face unique challenges in the patent process.[34] They may lack access to mentors and sponsors who can guide them, and they are more likely to encounter bias from patent staff and decision-makers. These obstacles, combined with poor communication and a lack of transparency, can make the patent process seem more daunting and inaccessible.

Ideas, whether generated through group collaboration or individual efforts, typically undergo review by patent attorneys. Some companies have in-house legal teams dedicated to intellectual property, while others outsource this function. For fast, internal innovation driven by a company's workforce, clear communication between lawyers and inventors is crucial. Time is a critical factor, as inventors often juggle multiple responsibilities and projects. They need to understand the steps involved, the expected timeframes, and the potential outcomes. Whether patent counsel is internal or external, standardized communication is essential in determining whether an inventor will continue to engage in the patent process.

Building a small piece of the metaverse or creating a virtual fantasy environment requires a vast array of expertise, spanning from ideation to final product development. As organizations and leaders race to build the metaverse, they cannot always predict who will have the solutions they need. Increasing the number of innovators means increasing the pool of potential solutions. Teams with diverse personal and professional backgrounds bring the agility necessary for innovation and product development. While not every idea will succeed, an unbiased idea assessment system is critical to closing the innovation gap and ensuring that all ideas are given a fair chance to flourish.

Barrier to Engagement

Many employees, even those at large tech companies, may not fully understand the metaverse or its potential impact on their work and careers. This knowledge gap can make it difficult to generate enthusiasm for metaverse-related projects and present challenges in engaging staff. Employees who see themselves as innovators might still believe that contributing to the metaverse requires advanced technical expertise. Companies cannot assume that all employees grasp the concept or significance of the metaverse. Continuous education and regular updates about the metaverse, including developments in VR and AR, are essential for keeping employees motivated to innovate.[35]

Innovation and processes like patenting can seem daunting and complex. Even with persistent encouragement from leadership, motivation can quickly wane, leaving employees fluctuating between inspiration and confusion. While innovation is always happening within the workforce, employees often hesitate to share their ideas. Effective communication can help move these ideas from mere concepts to actual product development. In this phase, peer-to-peer communication has proven particularly effective.[36]

Ambiguities can slow down an organization or defuse critical efforts, and the transition to the metaverse introduces exactly these kinds of challenges. Navigating them requires carefully planned, strategic campaigns that clarify goals and engage all internal stakeholders in innovation. Some organizations excel at crafting and sharing messages that inspire their people and

Designing for Impact 113

drive the business forward, while others become known for missteps that lead to employee departures and operational disruptions.

Under the pressure to develop new products and services for the metaverse, businesses must manage internal communications about innovation with intelligence and precision. Messaging needs to reach and include the entire workforce, empowering everyone to take action. Achieving this level of inclusivity can be challenging, but there are proven approaches that make it possible.

Diverse Voices Innovating

In the metaverse innovation workspace, diverse perspectives enhance both processes and outcomes. True innovation requires more than demographic variety—it thrives when businesses actively tap into contributions from all employees, regardless of background or role. Effective strategies not only increase participation from underrepresented innovators but also foster engagement across all teams, ensuring a culture where new ideas emerge organically.[37]

To encourage idea-sharing and sustained collaboration, companies must start with clear leadership communication that reinforces innovation as a collective effort. This should be followed by strategic internal messaging that maintains a continuous feedback loop, allowing employees to see the impact of their contributions. When organizations create an environment where participation is encouraged, valued, and acted upon, they naturally unlock new levels of creativity and problem-solving.

Teams with a range of expertise, not just demographic diversity, drive faster and more effective innovation. While diversity in race, gender, neurodiversity, and physical ability contributes to a richer workplace dynamic, diversity in skill sets and industry experience is equally critical. Cross-functional teams—bringing together technology, design, user experience, marketing, and finance—help companies navigate the complexities of metaverse development, reducing costly missteps and ensuring products are market-ready. In an innovation-driven, fast-paced digital economy, teams that leverage broad knowledge bases and adaptive thinking will be better positioned to deliver solutions that are both impactful and commercially viable.

Peer-Driven Innovation

Employees typically embrace the chance to support, inform, or guide their peers, as it enhances workplace culture and boosts morale. However, sustaining this initial excitement and integrating a sense of belonging into daily activities at both team and individual levels requires a well-crafted communication strategy.[38]

Reluctant innovators, particularly those who are less likely to see themselves as part of the innovation process, are more inclined to contribute when they can learn from others with shared experiences. Seeing innovators from similar backgrounds or professional paths provides a powerful sense of validation and encouragement, breaking down perceived barriers and fostering a culture where more individuals feel empowered to participate. Creating visible examples of diverse innovators helps organizations cultivate an environment where contributions feel both valued and attainable, ultimately strengthening the innovation pipeline.

Leaders play a crucial role in initiating this process but must also recognize when to pass the baton. They need to hand off the details of educating and engaging their teams to peers who are not only capable but also relatable. These peers can then move the process forward, ensuring that the campaign remains linked to the organization's purpose and resonates with all employees.

By empowering employees to lead by example and share their experiences, companies can create a continuous cycle of engagement that drives innovation across all levels of the organization. This approach not only sustains momentum but also helps build a culture where every employee feels valued and inspired to contribute their unique perspectives to the collective goal.

Peer-to-Peer Power

One effective way to reconfigure the communication process is to implement a peer-to-peer information system.[39] Best practices confirm that systems of informing and training are more sustainable when they are employee-led. Effective peer mentoring systems build awareness and prepare individuals to engage with a company's internal systems. Peer mentoring also supports idea development, optimizes the likelihood that the

Designing for Impact 115

message is received, and accelerates both innovation- and patenting-related decisions.

Peer instructors can walk inventors through the details and move them to the next level of action. This approach helps innovators understand and feel a sense of certainty about patents, defensive publication, and trade secrets. Peer mentoring systems support company goals related to inclusive innovation and increase the number of underrepresented innovators in the pipeline.

Research indicates that, given the choice, employees prefer to learn from other employees and that employees derive considerable workplace satisfaction from helping their colleagues.[40] In order to be successful, peer mentors need only to thrive at connecting with others and ensuring that they have what they need to succeed. Over a dozen companies proudly declare that "innovation is in our DNA." Sometimes those genes can remain unexpressed. Peer mentoring can help activate this potential.

One-on-one contact with a peer who is more knowledgeable can be used in conjunction with an examination of other company data to foster a sense of belonging and help drive inclusive innovation. Peer-to-peer interaction and education prevents some of the communication pitfalls that stymie organizations, especially when there is an urgent need to innovate inclusively and successfully. Many people are comfortable with creative processes and problem-solving; however, after devising an idea or developing prototypes or explanations of the related function and novelty, many innovators hit an impasse. The next steps may seem unclear or intimidating. The processes associated with patenting and IP in general still seem arcane to most employees. And when it comes to the metaverse, patents remain a major unknown. Uncertainty or stress can prompt individuals to disengage or self-exclude. Access to informed peers can mitigate this stress and foster innovation. In the context of communication, the peer-to-peer approach increases the likelihood that the message will be received and acted upon. Access to the messenger makes follow-up and deeper learning more likely.

Empowering Peer Educators

Organizations should ideally build a team of peer educators using individuals who have a passion for championing the patent process and the organization's resources. A key element of their role and mission is to spark interest and action in others. Helping their peers initiate and navigate the patent process at their organization requires the kind of practical familiarity that lets the peer educator serve as a significant source of accurate information. Peer educators don't need to personally have an issued patent or be from any specific area of the company. The capacity to connect personally and a passion for making sure that others have access to the right people and resources are core characteristics. These individuals are more likely to reach employees who may be reluctant to engage. Many already offer outreach, information, assistance, guidance, and other support on an informal basis or in a supervisory capacity. Building a formal process is essential to tuning the message so that the information flowing horizontally from peer to peer resonates with the vertical message from leadership. In addition to consistency, education, and engagement, structuring the process provides other benefits.

Effective trainers maintain consistent, impactful communication from leadership to the workforce. In order to motivate employees and prompt them to take action, messages from leadership must be extended to everyone in a manner they can practice. The representation of peers as inventors removes barriers and triggers self-inclusion. Examples of innovators who reflect the diversity of the firm can be captured and showcased using short videos and other media. These tools should candidly showcase an inclusive selection of innovators at an organization. Seeing others like themselves reflecting on their experiences with the patent process creates psychological safety and incentivizes engagement for employees.

Centralizing innovation resources on an organizational webpage or hub offers a one-stop source for innovators. The venue extends the communication effort and helps the business clarify what it is seeking, from whom, and why. Engaging everyone equally ensures that innovation is inclusive. Diverse problem-solvers bring the unique perspectives formed by their individual backgrounds and experiences. Unique problems, gaps, and difficulties may drive responsive innovation.

The Prize Is . . .

The need to drive innovation as businesses plan to enter the metaverse or seek to thrive there may be obvious, but it is important for businesses to communicate the problems they are trying to address and explain why they are seeking to innovate to their employees. Leaders should craft clear and concrete calls to action. Both the peer instruction process and the innovation webpage should nudge visitors to share their ideas and inform them about what they need to do to submit ideas. Once individuals are motivated, they may want to learn more or may have further questions. As part of the horizontal communication around patents and innovation in general, businesses should have clear, consistent policies in place, as well as answers to potential questions.

Investing in a group of individuals whose temperaments, experience, and interests make them ideally suited to serve as peer mentors and advisers to fledgling innovators facilitates the horizontal aspects of the communication campaign. Presentations, instructions, encouragement, and advice delivered by peers help flesh out the mandate of leadership at the level where innovation most often happens. Such communication integrates the primary inspiration and excitement into a culture of daily actions that builds the big picture. Leaders have an outstanding opportunity to organize and implement peer mentoring systems that remove barriers, distribute information, and positively shift the number of people who are aware of innovation resources and participating in the process. In this way, more innovators will be getting their ideas out there and the diversity of voices will continue to grow.

7 Leveraging Strategic Alliances

A key skill in technology and innovation is the ability to forge successful collaborative partnerships.[1] The right partnership can accelerate innovation, streamline development, and make commercialization more feasible. Companies are increasingly joining forces to develop products and services they couldn't create alone, reach new markets, reduce costs, and gain access to cutting-edge technologies and expertise.

Many of the innovative products we use today are the result of such collaborations. For example, the Blu-ray disc was born from the partnership of Sony, Philips, and Panasonic, while USB technology came from the combined efforts of Intel, Compaq, and Microsoft. Ethernet, developed by Digital Equipment Corporation (DEC), Xerox, and Intel, has been in use since 1980. Similarly, technologies like Wi-Fi and GPS also arose from successful partnerships, demonstrating that technological breakthroughs often require a broad range of talent and expertise.

In the emerging metaverse, partnerships continue to drive innovation. Google and Meta are collaborating to bring Google Workspace to Meta Quest Pro headsets, enabling users to access Gmail, Google Calendar, and Google Docs from their headsets.[2] This marks an early step in metaverse integration, much like the progression from the original iPhone to today's models, which consolidated multiple devices into a single, powerful tool. Future innovations will likely further miniaturize and integrate these technologies, making them less conspicuous and more universally accepted in everyday life.

Meanwhile, other collaborations are shaping the future of the metaverse. Microsoft and Samsung are developing business-oriented Windows laptops and tablets.[3] Amazon and Intel are optimizing cloud computing services for artificial intelligence (AI) and machine learning workloads.[4] IBM and the software company Red Hat (now acquired by IBM for 34 billion USD) are working on a hybrid cloud platform to help businesses manage their workloads across on-premises and cloud environments.[5] Nvidia and VMware are creating a high-performance gaming platform for the cloud.[6] These partnerships exemplify the new normal in technological innovation, many of which will have significant impacts on the metaverse.

Additionally, the Meta–Microsoft collaboration will allow users to access Microsoft Teams from Meta Quest Pro headsets, facilitating collaboration within the metaverse.[7] Epic Games and Sony are developing new metaverse experiences for PlayStation users, while Nvidia and Unity are making it easier for developers to create and deliver metaverse content.[8] Google Cloud and Roblox are partnering to help developers create more scalable and reliable metaverse experiences, and the partnership between Hewlett Packard Enterprise (HPE) and Microsoft aims to leverage HPE's infrastructure with Microsoft's Azure cloud platform to support enterprise metaverse initiatives.[9] FedEx and Microsoft will also solidify their cross-platform "logistics-as-a-service" solution.[10]

Why Partnerships Became the Norm

Constant change and rapid, unpredictable obsolescence are real dangers in metaverse-related industries, making it clear that one is still the loneliest number. Working with a partner significantly reduces the risk of failing to innovate, experiencing a short product lifecycle, or facing outright consumer rejection.[11] Partnerships offer unique opportunities for diversity and inclusivity, bringing together diverse perspectives, talent, workstyles, and resources.

As businesses across industries recognize the value of collaboration, partnerships are becoming the norm, with 94 percent of tech executives viewing them as a critical strategic element.[12] These collaborations offer more than just extended expertise—they can transform companies into powerhouses of innovation and breakthroughs. For businesses aiming to make a mark in the metaverse, collaboration may be essential. Going solo may not be feasible when faced with major technical challenges that exceed an organization's capabilities. Partnerships provide a competitive edge, allowing companies to overcome obstacles and outpace their rivals. In the global metaverse economy, partnering enables companies to expand their reach and enter new markets, proving that partnered businesses truly represent more than the sum of their parts.

Starting with Clear Goals

Partnerships, while offering significant advantages, are fraught with potential pitfalls. Without clear goals, measuring progress and making necessary adjustments becomes challenging, and issues may go undetected until it's too late. For example, Apple and Google's 2006 partnership to develop what became the Android operating system highlighted a clash in visions—Apple sought a closed, proprietary system, while Google aimed for an open platform accessible to other manufacturers.[13] Misaligned goals, workstyles, market share considerations, and regulatory constraints can all derail partnerships.

Not all connections are a good fit—individual goals may not align, mutual goals may seem unattainable, and communication can fail.[14] While innovating with a partner can lead to accelerated innovation and improved efficiency, it also poses risks. Trust is crucial, as collaboration can expose each partner's intellectual property, aspirations, and strategic plans.

Effective strategies for forming and managing innovation partnerships are essential to making these collaborations successful.[15] Companies enter partnerships with a set of hopes, but it's crucial to transform these hopes into clear, agreed-upon expectations and goals. Establishing concrete objectives, roles, and timelines can help ensure that partnerships remain focused and productive, increasing the likelihood of achieving the desired innovative outcomes.

Trust and Communication

Trust and communication are as vital to strategic partnerships as they are to personal relationships.[16] For partnered companies to succeed, they must trust each other to share information, collaborate effectively, and make mutually beneficial decisions. Given that strategic partnerships are often long-term, clarity and candor at the outset are essential to prevent future conflicts or disappointments. Partners need to establish ways to be open and forthcoming; without transparent disclosure, they cannot serve each other effectively.

Trust in partnerships extends beyond NDAs and other formal agreements.[17] Historical examples underscore this point: the DaimlerChrysler merger, one of the largest in history, failed due to unrealistic expectations, cultural differences, and a lack of trust.[18] The Sony–Ericsson joint venture was successful for several years but ultimately faltered due to differing goals and poor communication.[19] The once-prominent Nokia–Microsoft partnership also crumbled under the weight of misaligned goals and unrealistic expectations.[20]

Disruptive factors can easily unravel even the most ambitious partnerships. When one partner is treated as ancillary or merely a means to an end, they may become more focused on self-preservation and less motivated to protect the interests of the more dominant partner. Conversely, partners who view each other as equals are more likely to safeguard each other's intellectual property and interests. This foundation of trust and mutual respect is crucial for building and adapting long-term strategic partnerships.

However, the complexities of these relationships, coupled with the urgency of building the metaverse, may amplify tensions. Just as personal relationships require nurturing, strategic partnerships need attention to trust, communication, and mutual respect from the start.[21] By building these elements into partnerships up front, companies can navigate the complexities and sustain successful collaborations.

Responsibilities and Communication

The distribution of responsibilities and effective communication are critical to the success of any business-to-business partnership.[22] Many collaborative efforts fail due to a lack of clear role definitions and performance expectations. While partners may unite around a shared goal or innovation, the specific roles and contributions of each party are often not clearly outlined. This ambiguity increases the likelihood of disputes, diverting time, effort, and resources away from innovation. Clearly defined responsibilities and expectations may not eliminate tension or conflict entirely, but they significantly reduce the chances of disputes escalating.

For a partnership to be sustainable, communication is key. It serves as the foundation for maintaining trust and transparency. A failure to disclose or clarify issues can lead to misunderstandings, ultimately ensuring that neither partner benefits fully. Regular communication channels, such as weekly meetings or daily check-ins, are essential for identifying and addressing potential issues early. In situations of uncertainty, the default should be full disclosure. Additionally, training employees at both companies on effective communication practices can help prevent disputes from arising, minimizing the need for formal dispute resolution measures.

Money

In collaborative partnerships, money plays a role similar to its function in personal relationships: it can be a source of motivation and sustenance or an obstacle that leads to conflict and dissolution. This is especially true in partnerships between large companies and smaller partners, such as vendors or suppliers, where financial resources and contributions are often unequal.[23] If financial matters aren't resolved early on, they can breed stress, conflict, and resentment, as seen in the Sony–Ericsson and Nokia–Microsoft partnerships, where money-related issues contributed to their eventual failure.

In strategic partnerships, money is involved in two key ways: "money in," which refers to the investment required from all partners, and "money out," which includes revenue and other monetizable benefits. In this context, money is closely tied to risk.[24] Partners are motivated by shared needs and opportunities, and it's crucial that they craft agreements that are fair and sustainable for all parties involved. If these financial agreements are not equitable and advantageous, the partnership is likely to fail. Successful collaborations are built on a balanced distribution of risk, underpinned by trust. By pooling resources and expertise, partnerships can accelerate innovation, creating wins for businesses, users, and the broader economy.

However, collaborations are not meant to last forever, making the equivalent of a prenup often necessary. Successful partnerships may generate innovations beyond the initial scope, raising questions about who gets to use and benefit from the resulting intellectual property (IP). If a party came into the collaboration with specific technical outcomes in mind, and these were achieved, it's clear who should own the resulting IP. But when one party has no use for certain IP, allowing the other party to benefit can make sense and foster goodwill. Protocols for IP allocation should align with the partners' business interests through appropriate filings and licenses, ensuring that the IP generated benefits all involved rather than merely expanding one partner's portfolio of unused patents.

Dispute Resolution

Money, ownership, management, milestones, and IP are common sources of disputes in strategic partnerships. When these issues aren't addressed through timely and effective communication, tensions can escalate, often leading to the activation of dispute resolution mechanisms. It's crucial that this system be established by equally empowered representatives from both partner companies, regardless of their size, to avoid any perception of bias. The goal is to foster more innovation than contention. Ideally, conflicts should be resolved at the point of disagreement, with escalation up the management chain reserved only for situations where it is absolutely necessary. Early resolution through negotiation, communication, clarification, and mediation is key to maintaining a productive partnership.[25] Arbitration and litigation should be considered last resorts, as they often signal the end of a partnership. External arbitration in particular can disrupt or even terminate a collaboration, typically resulting in more losses than gains for both parties involved.

The Payoff

Successful partnerships offer significant payoffs. For example, the Nike+iPod sports watch led to a surge in fitness tracking devices, and collaborations like Xbox Live between Microsoft and Sony have been mutually beneficial. Tesla and Panasonic's joint efforts produced advanced electric car batteries, while entertainment industry partnerships such as Netflix and Marvel, and Disney and Pixar, have also seen great success. In the metaverse economy, collaborations between companies like Google and Unity, Nvidia and Epic Games, and partnerships involving Decentraland, Samsung, Warner Bros., and others are laying the groundwork for future innovations.

Prioritizing the key elements of successful partnerships—trust, clear expectations, shared risks and rewards, and effective conflict resolution—sets the stage for fruitful collaboration. Approaching potential partners with a flexible draft of needs and possibilities fosters trust and interest. Pooling resources and talents can create a powerful engine for innovation, accelerating the development of the metaverse and benefiting a broader user base.

A systematic approach to sharing risks, reaping benefits, and resolving differences is essential in any collaborative partnership. Financial risks must be managed carefully to ensure all parties feel secure throughout the collaboration. Establishing optimal protocols for IP ownership and commercialization rights, as well as equitable conflict resolution methods, ensures that both partners can thrive. Leaders should recognize that addressing trust, expectations, risk, reward, and conflict resolution up front can lead to successful innovation partnerships. One significant payoff of partnering is the infusion of diverse perspectives, which can lead to smoother product rollouts and broader adoption.[26]

Customers and Innovation

In addition to engaging in partnerships, another beneficial and well-established practice for many businesses is integrating customers into the innovation process. Companies like Intuit, Amazon, Tesla, and Google regularly gather customer feedback through advisory boards, user testing, and beta programs to shape their products and services.[27] Even Apple, known for its secrecy, leverages customer input through its support team, and NASA, one of the premier scientific organizations in the US, engages customers through its citizen science program to enhance innovation. Across industries, businesses that involve customers in the development process report increased customer satisfaction, faster time to market, and higher innovation success rates.[28] Early customer feedback not only helps identify and resolve potential issues but also builds inclusivity into products and services from the start.[29]

Innovating inclusively is increasingly recognized as essential for successful product launches, yet achieving this remains challenging. Many industries are experiencing diversity fatigue, with some companies struggling to maintain progress, especially in the months following the June 2023 decision from the United States Supreme Court that struck down affirmative action in university admissions.[30] Tech companies, in particular, often lack sufficient diversity to provide the necessary inclusive input, especially in areas like AI development. However, the need for diverse feedback is clear, as it can prevent potential backlash and ensure products meet the needs of a broader audience.[31]

Businesses should actively seek input from existing and potential customers to understand how the metaverse will be used and how much time people will spend in augmented reality (AR) and virtual reality (VR).

Customers can provide valuable insights into their expectations and usage patterns, guiding companies to build a metaverse that exceeds their expectations. Recent surveys, such as one by Statista, indicate strong consumer interest in VR headsets, with 56 percent of respondents expressing interest in purchasing one within the next two years.[32] Among younger demographics, particularly those aged 18–29, the figure rises to 63 percent. However, barriers like limited content and high prices remain.

Businesses are increasingly using surveys, interviews, and focus groups to gather inclusive feedback from customers. Customer advisory boards, diverse in composition, offer ongoing input, while social media monitoring and analysis of customer support data help identify patterns of discrimination or bias. Proactively engaging customers helps companies innovate inclusively and avoid legal challenges and reputational damage.[33]

Examples of legal actions, such as lawsuits against VRChat, AltspaceVR, Somnium Space, and Meta Horizon Worlds, highlight the consequences of failing to incorporate inclusive features.[34] Research by the Pew Research Center shows that AR and VR filters can exacerbate issues like racial bias, anxiety, and depression, particularly among marginalized users. A study by a team from the National Institutes of Health in Bethesda, Maryland, found that Black and Hispanic users are less likely to be interested in using AR or VR due to concerns about discrimination.[35] Researchers also discovered that AR filters that change a user's appearance are more likely to be used by white users than by Black users, a trend attributed to racial bias in the filters. Moreover, research indicates that these filters can increase feelings of anxiety and depression, which could be particularly harmful to already marginalized users.

By engaging customers early and often, companies can create more inclusive, successful products and avoid the pitfalls that come with exclusion.

The Need for Customer Feedback

The customer collective often holds insights that companies may overlook or fail to fully articulate. Apple famously used a focus group to shape the design of the first iPhone, and other giants like Nike, Amazon, Netflix, and Tesla are known for integrating customer feedback into their innovation processes.

As they build for the metaverse, companies like Meta, Microsoft, and Nvidia are relying heavily on customer insights.[36] For instance, Meta used feedback to introduce custom avatars in Horizon Worlds, while Microsoft added real-time language translation to its Mesh platform based on user requests. Nvidia's customer feedback led to the development of new metaverse hardware and software, including the Omniverse platform.

Even the most successful companies need customer feedback to make the metaverse welcoming and sustainable for everyone. This input helps identify biases, informs avatar design, and shapes interactions in virtual spaces. For example, Meta is using customer insights to prevent harassment and discrimination in Horizon Worlds, while Nvidia is developing hardware that enhances avatar realism and expressiveness.

As businesses continue to build their segments of the metaverse, they have a valuable opportunity to engage with customers as consultants.[37] This collaboration can yield transformative insights that guide optimal, sustainable decisions and help ensure that new technologies are widely adopted.[38] While maintaining strategic secrecy may be necessary, companies can still gather critical feedback from potential customers to prevent design missteps and market failures.

Engaging customers early fosters trust and loyalty, leading to increased sales and long-term success. Feedback helps businesses prioritize innovations, predict product viability, and optimize resources, reducing the risk of sunk costs. With the rising adoption of metaverse technologies, as evidenced by the projected increase in sales of augmented reality glasses and VR headsets, engaging customers to meet evolving expectations is more important than ever.

The number of people entering the metaverse is rapidly increasing as technologies advance and prices decline. Statista projected a dramatic increase in consumer augmented reality glasses sales—from 10,000 units in 2019 to 1.59 million by 2024—signaling a strong upward trajectory in adoption.[39] This surge will require companies to actively engage with customers to better prepare and expand the metaverse in line with their expectations and needs. Similarly, VR headset sales were expected to grow from 6.1 million units in 2021 to 16.44 million by 2024. The introduction of devices like Apple's Vision Pro and Meta's Quest 3 in 2024 was positioned to further accelerate consumer interest.[40] As 2025 unfolds, early market data will likely confirm whether these projections have held—and whether immersive technologies are reaching an inflection point in mainstream adoption. If current trends continue, these technologies may soon shift from early adopter enthusiasm to widespread consumer integration, much like the smartphone and streaming revolutions of earlier decades.

Lessons from Google Glass

Google Glass offers a compelling case study in the consequences of overlooking or disregarding behavioral science insights. Launched with much excitement in 2013, these "smart glasses" promised a hands-free, AR-like experience, but despite initial interest, they encountered significant consumer resistance.[41] By 2023, Google had discontinued the product, marking it as a notable failure. The first lesson here is that even technologies with clear benefits can struggle with consumer adoption.

The technology itself wasn't the issue; rather, several behavioral factors contributed to its downfall. At $1,500, the price was steep. The battery life was disappointingly short, requiring a recharge after just four hours. Far from achieving the desired cool aesthetic, the glasses were widely ridiculed for their awkward appearance. This leads to lesson two: new technologies must be designed with the user in mind.

Furthermore, Google Glass raised concerns about privacy—it allowed users to easily record others without their knowledge, raising legal and ethical questions. Safety concerns also emerged, such as the potential for users to become distracted by the display and walk into traffic. Most importantly, consumers questioned the need for Google Glass when their smartphones already provided the same functionality. This brings us to lesson three: it is crucial to consider the social and ethical implications of new technologies.

It's unclear whether behavioral science was employed in the development of Google Glass, but several aspects could have been tested: acceptable price points, consumer tolerance for battery life, reactions to the design, privacy concerns, safety risks, and overall desirability. Behavioral scientists specializing in innovation might have challenged the developers' assumptions, potentially leading to significant modifications or even the decision to abandon the product altogether.

Unique Technology

Metaverse technology stands apart from traditional tech like laptops or smartphones due to its deeply immersive and personal nature. Unlike other devices, metaverse-related technology is worn, becoming an extension of the wearer and influencing both their self-perception and how others perceive them. This unique characteristic introduces a range of behavioral responses that can significantly impact a product's success or failure.

To leverage this factor effectively, it's essential to integrate design elements that consider these behavioral dynamics. Behavioral science offers a way to move beyond mere guesswork, providing measurable insights into how people interact with and react to new products. Many failed innovations, such as Google Glass, serve as reminders that consumer rejection often stems from how a product looks, feels, or is perceived by others. Understanding and addressing these behavioral aspects can make the difference between a product being embraced or relegated to the failed-product graveyard.[42]

The Hazing of Oculus

One product is facing those particular challenges today. Oculus, despite its relatively recent entry into the market, has faced a difficult journey marked by technical issues, content shortages, competition, pricing, and privacy concerns. The developers initially raised $2.5 million through Kickstarter in 2012, and just two years later, Facebook acquired the company for $2.3 billion—a move that sparked controversy due to Facebook's privacy practices. Oculus's first major release, the Rift headset in 2016, was both a critical and commercial success, though it came with a high price tag and some technical issues. The 2018 release of the Oculus Go offered a more affordable and portable VR experience but was less powerful and had limited content. In 2019, the standalone Oculus Quest headset eliminated the need for a PC or console, achieving commercial success despite its high price and limited content. The release of the Quest 2 and Quest 3 improved accessibility by being more powerful and affordable, marking a significant advancement in VR technology.[43]

New technologies like Oculus often endure a rough start before gaining widespread acceptance. The path to success can be brutal, with consumers and reviewers quick to criticize. On Facebook, reviews of Meta's Oculus headset are a mix of high praise and harsh criticism, focusing on performance, comfort, price, and aesthetics, alongside concerns about needing a Facebook account and privacy issues. Meta has responded to these critiques, with the Quest 3 offering improved resolution, ergonomics, and performance.[44] To address cost concerns, companies are working to improve manufacturing efficiency, reduce costs at the component level, and offer a range of headset models at different price points.

Designers are also addressing the aesthetic and comfort aspects of VR headsets. By creating sleeker, slimmer models with reduced size and weight, and incorporating adjustable headbands and better cushioning, manufacturers aim to enhance user comfort. High-quality materials, customization options, and collaborations with high-profile fashion brands are further nudging consumer behavior toward wider adoption of VR technology.

Innovating for Accessibility and Engagement

Creating sleek, user-friendly headsets is just the beginning—true innovation in the metaverse requires a deep understanding of diverse user needs. Just as iPhones and Androids or Macs and PCs serve different user preferences, AR, VR, and MR (mixed reality) technologies must accommodate a wide range of activities, devices, and user experiences. Inclusive design must consider not only what people want to do in the metaverse but also who is using these technologies. Products and services should be accessible across various devices and browsers, support multiple languages, and incorporate features that accommodate individuals with disabilities. Moreover, ensuring fairness in policy design, infrastructure expansion, education, and information dissemination is key to making the metaverse's economic opportunities widely accessible.

Every stage of the commercialization pipeline—idea generation, innovation, product development, sales, and marketing—involves human behavior. Behavioral science offers valuable insights into optimizing these processes by identifying and minimizing biases that unintentionally exclude users. A metaverse that authentically reflects global diversity is not just idealistic—it is essential for long-term economic success and user adoption.

Larger companies may have the resources to recover from missteps, but smaller businesses and startups do not have that luxury. Many organizations and governments are turning to behavioral science teams to guide decision-making, improve user engagement, and mitigate biases that limit innovation and market reach. By integrating psychology, economics, and strategic design, businesses can navigate uncertainties in the metaverse, enhance user experiences, and build sustainable digital ecosystems.

Early Warning and Better Guidance

Smart companies integrate customer feedback early in their innovation processes through focus groups, surveys, and other input methods. This approach can significantly reduce the likelihood of product rejection. For instance, Microsoft's Kin phone, designed for social networking, failed within two months of launch because it overlooked a crucial feature: the ability to post pictures to Twitter, a key platform for its target audience. This oversight highlights the dangers of innovation conducted in an echo chamber, where leaders and employees brainstorm ideas without meaningful customer input, leaving success up to chance. This internal echo chamber can span departments, roles, and even continents, introducing significant risk to product development and marketing. A large, multicontinental enterprise might appear formidable in strategy, but without the insights, preferences, and behaviors of potential buyers, it gambles with its resources. In the fast-paced metaverse economy, where technological developments come at high costs, businesses must balance innovation with calculated risks. While adopting a safe-to-fail approach encourages experimentation and learning, companies must also implement strategies to mitigate large-scale failures that could jeopardize financial stability or market position. Leaders must focus on adoption, as seen in Meta's ongoing partnership with EssilorLuxottica (Ray-Ban) to ensure its smart glasses are not just technologically advanced but also "cool" and appealing to consumers. Psychologically, "cool" and "unique" stand in stark contrast to "nefarious"; customers make that distinction.[45]

By listening to customers and incorporating their feedback, companies can avoid failures like Google Glass. A smart approach to giving people what they want increases the likelihood of product adoption. Customer feedback can be accessed through direct input and big data, guiding businesses away from viewing inclusion as a mere corporate obligation. In the metaverse, inclusive design is not a corporate cliché but rather a critical strategy for success. Businesses that get it right will win early and continue to succeed. Effective customer partnerships help identify potential flaws, choose optimal price points, and ensure that innovation and development are guided by real-world insights, not guesswork.

III Metaversal Frontiers: Transforming Industries and Professions

8 The Public Sector: Government, Education, and Health Care

Across the globe, governments and public sector organizations are stepping into the metaverse, harnessing its potential to enhance public services, foster global cooperation, and engage with citizens in innovative ways. Major cities are building virtual replicas, the United Nations is hosting virtual conferences, and countries like Estonia and Barbados are pioneering virtual embassies. These efforts signal a new era of digital governance, in which the metaverse becomes a critical platform for urban planning, public service delivery, and international diplomacy. Organizations such as the Red Cross, the World Health Organization, Amnesty International, the World Wildlife Fund, and Médecins Sans Frontières (Doctors Without Borders) are utilizing virtual reality (VR) to raise awareness and deliver support. As governments and public sector agencies integrate into this evolving digital landscape, they are not only expanding their reach but also reimagining how they serve and interact with the public, setting the stage for a more connected and responsive world.

The Public Sector

In the real world, the public sector often goes unnoticed, yet it provides essential services like education, health care, and infrastructure, while regulating the private sector. The metaverse has the potential to revolutionize how the public sector plans, assesses, executes, and maintains projects, as well as how it delivers services.[1] For example, the US Department of Transportation is already utilizing the metaverse to create digital twins of cities and infrastructure, enabling advanced planning and simulation for transportation systems and disaster response.

Governments should consider integrating into the metaverse because it offers a platform where they can create detailed virtual replicas of cities, landmarks, and historical sites, enhancing tourism appeal and cultural outreach.[2] Cities like Seoul, New York, London, and Dubai are already leveraging the metaverse to showcase a seamless blend of real-life governance and virtual possibilities.

In Santa Monica, California, a partnership with the metaverse platform FlickPlay has created a unique play-to-earn application that combines social media, gamification, VR, and augmented reality (AR).[3] This innovative approach enriches social interactions by embedding digital layers across city spaces, allowing citizens to gather virtual collectibles and earn real-world rewards, demonstrating how digital innovation can fuel community engagement.

These virtual platforms not only offer impressive marketing opportunities for tourists and investors but also provide innovative approaches to urban planning and public service strategies. Remote public services are becoming more popular, with cities using the metaverse to allow citizens to apply for passports without visiting government offices. Public safety, first responders, and economic development initiatives are also entering the metaverse.

Hong Kong's MTR Corporation Limited, for instance, has entered the metaverse through a strategic alliance with The Sandbox.[4] MTR is creating a virtual railway ecosystem with stations and a museum designed for interactive, gamified citizen experiences. By purchasing virtual land, MTR is not escaping reality but enhancing it, using the metaverse to strengthen community ties and reflect its physical services in the digital realm.

As the metaverse expands, it challenges traditional governance to imagine a world where digital sovereignty complements physical jurisdictions.

Who's in First?

The Republic of Korea and its Ministry of Science and ICT (Information and Communications Technology) is making a significant investment of at least $186.7 million to develop its metaverse ecosystem, with the potential to create 1.5 million job opportunities. As part of this strategy, the nation plans to employ over 40,000 professionals specializing in metaverse development.[5] The government's objectives are to activate metaverse platform ecosystems, develop skilled professionals, foster corporate growth, and ensure a safe user environment within the metaverse, with a focus on applications in arts, culture, education, K-pop, and tourism.

To support these goals, plans are underway to establish a metaverse academy and organize development contests to encourage creative culture and participation from citizens and businesses. The government will also introduce a new metaverse hub, providing space, facilities, and financial backing through a dedicated metaverse fund to support startups. In collaboration, the Electronics and Telecommunications Research Institute (ETRI) and the Korea Information and Communication Technology Industry Association are set to develop an ICT convergence standard framework for the metaverse.[6] Alongside technological advancement, the government aims to create ethical guidelines and will form a pan-governmental entity to review and adapt relevant laws and regulations. The Korea Information Society Development Institute has outlined ten principles for future-oriented regulations of metaverse platforms, advocating for a case-by-case approach to align with global regulatory harmonization efforts.

Seoul, the capital of Korea, announced a five-year blueprint in November 2021 to create Metaverse Seoul, aiming to be the first city replicated in the metaverse.[7] This plan spans seven domains: economy, education, tourism, communication, city, administration, and infrastructure. The five-year plan unfolds in three stages.

The introduction phase of Metaverse Seoul began in 2022 and focused on establishing the platform and introducing services across key sectors such as the economy, education, tourism, public services, and urban management. This phase culminated in a public presentation of the completed platform. Moving into the expansion phase from 2023 to 2024, the plan included setting up a virtual general civil service office, allowing citizens to interact with public official avatars to resolve civil complaints and access consultancy services without the need to visit city hall in person. Additionally, virtual recreations of Seoul's famous tourist destinations were integrated into the platform to enhance user engagement. The final settlement phase in 2025 and 2026 aims to further enhance and refine these services, solidifying Metaverse Seoul as a comprehensive digital twin of the city.

Pioneering Diplomacy?

Barbados is boldly extending its sovereignty into the metaverse, charting an unprecedented course in diplomatic innovation.[8] In a landmark initiative, the nation's Ministry of Foreign Affairs and Foreign Trade is collaborating with Decentraland to establish the world's first metaverse embassy. This groundbreaking move redefines diplomatic norms, offering a cost-efficient and impactful alternative to traditional embassies, particularly for nations with limited resources.[9] Ambassador Gabriel Abed, representing Barbados in the United Arab Emirates, highlights the significance of this venture, envisioning metaverse embassies as gateways to diplomatic equality on the global stage.

A virtual embassy goes beyond a mere digital representation of a nation; it's an immersive space where diplomacy, culture, and national identity converge. This platform offers interactive experiences for diplomats, investors, and the global community, promoting Barbados' rich heritage and tourist attractions and facilitating investment dialogues. It also enables Barbados to conduct a wide range of diplomatic activities seamlessly, engaging with nonresident ambassadors and international partners without geographical constraints.

This paradigm shift moves diplomatic interactions from physical offices into a limitless virtual environment, signaling more than just technological advancement but also a strategic step toward a future where digital sovereignty becomes a cornerstone of international relations. As Barbados leads the way, it invites the world to witness the evolution of diplomacy within the infinite dimensions of the metaverse. The question now is: How will other nations respond to this digital revolution?

Service, Service, Service

The metaverse, enhanced by artificial intelligence (AI) that enables customized interactions, offers government agencies the potential to significantly upgrade their customer service capabilities. Multilingual avatars can recall previous encounters flawlessly, access multiple databases, and provide effective, interactive experiences. Free from the constraints of time and location, metaverse-based government platforms can streamline interagency actions, allowing for faster collaboration and resolution of issues that might be intractable in the real world. The evolution from traditional internet services to the metaverse promises to further reduce wait times and expand the dimensions of public service delivery. However, this transition requires careful consideration of security.

The robust security features of blockchain technology are crucial as governments venture into the metaverse, carrying vast amounts of sensitive data.[10] Governments manage an extensive array of services and hold enormous stores of individual and business information, including financial records, biometric data, and personal details of millions—or even billions—of citizens. The potential for data breaches is a significant concern, as past incidents have shown. For example, breaches in the US have affected millions, such as the 2011 State of Texas breach, which compromised the data of 3.5 million people, the 2012 South Carolina Department of Revenue breach, which affected 3.6 million people, the 2009 hack at the National Archives and Records Administration (NARA), impacting 76 million people, and the 2015 US Voter Database breach, which impacted 191 million people.[11]

As governments integrate more services into the metaverse, the secure nature of blockchain technology can help verify transactions and protect privacy, especially in handling sensitive documents. Unlike the traditional internet, the metaverse offers more reliable methods for confirming identities, including the use of biometric data, which demands robust security measures. With the growing amount of data at stake, privacy and security have become paramount. Although the first major metaverse data breach has yet to occur, the potential risk is significant. Governments must proactively determine what information is collected and ensure it is rigorously protected to prevent breaches and safeguard citizens' trust.

Testing and Training

Governments can leverage the metaverse not just for environmental and infrastructure projects but also to test government operations and policy impacts. VR allows for the safe piloting of structural changes, such as municipal leadership models, before implementing them in the real world.

This factor of safety extends into the realm of training, where the metaverse offers an immersive environment for rigorous, specialized training programs. First responders and law enforcement personnel, for instance, can engage in scenario-based training, allowing them to respond to various emergencies or threats effectively. This kind of VR training expands skillsets and ensures appropriate responses to different levels of emergencies.

Military Applications

The US military is increasingly integrating the metaverse into its operations, recognizing its potential across a broad range of defense-related activities. The metaverse's capacity for digital twinning offers significant advantages, enabling cost-effective design, testing, and development of military equipment through virtual simulations. This technology streamlines engineering and production processes and enhances national defense by providing realistic, immersive training environments.

The Department of Defense is already leveraging AR and VR for training purposes.[12] For example, pilots have conducted simulated refueling maneuvers using virtual fuel tankers, highlighting the metaverse's potential to safely replicate complex, high-risk scenarios.[13] As the military continues to develop synthetic training environments that blend live and virtual elements, soldiers can engage in simulated battle scenes and outdoor exercises, navigating real-world terrains while interacting with virtual opponents. These innovations ensure that military personnel receive comprehensive, effective training that prepares them for a wide range of real-world challenges.

Strategic Integration and Governance

To successfully integrate into the metaverse, governments and public sector organizations must meet essential technological requirements, such as upgrading their hardware infrastructure and data systems. Building or rebuilding applications to include APIs and microservice architectures will be crucial for ease of use and sharing.[14] Additionally, governments will need to hire experts, including game designers, 3D artists, programmers, and network specialists, to facilitate and sustain their metaverse presence.

Strategic alliances will also be essential for sharing best practices and learning from others' experiences. Governments must be proactive in crafting policies related to data and technology use and in communicating their guidelines clearly to users. As they enter the metaverse, governments should not delegate their presence to other parties; instead, they should maintain control over their metaverse activities, especially since they deliver crucial customer-facing services and engage in significant research and development.[15]

Moreover, the metaverse will require regulation, and it will likely fall to governments to address how crime, civil law, and intellectual property rights will operate in this new environment. The decentralization inherent in the metaverse presents challenges, as it may remove certain activities from the traditional legal frameworks provided by national, state, or local governments. As regulators watch and prepare to act, governments must start grappling with these complex issues.[16]

The Critical Role of Government

As the metaverse continues to take shape, the public sector's expertise will be critical. While governments may initially take a light-touch approach, the need for rules, policies, and context will become evident as events unfold and people interact within this new digital realm. Legislation for the metaverse cannot be crafted in isolation; input from businesses and users will be essential to understanding the full scope of its needs. The role of government in the metaverse will be continuous and evolving. Both governmental and private sector leaders recognize that the metaverse will require some form of basic law and order, with governments and their regulatory resources naturally playing a vital role in maintaining structure within the emerging worlds of AR and VR.

Governments cannot afford to take a passive approach to the metaverse. Now is the time for them to begin discussing and shaping its future. Their priorities and perspectives are crucial as technology companies and users wrestle with ideas related to governance. While the metaverse is the result of long-term private sector innovation, its sustainability will depend on the policies, laws, and regulations that only governments can provide.

Education

While governments will help shape the metaverse, one of the things that the metaverse will shape is the future of education. The World Economic Forum's Education 4.0 Alliance, known as the Reskilling Revolution, emphasizes key skills essential for the future economy: creativity, analytical thinking, digital literacy, AI, and collaboration.[17] The metaverse, with its interactive environments, is ideally positioned to teach these skills. Jensen Huang, CEO of Nvidia, has pointed out how new AI tools are lowering barriers to learning programming, demonstrating the significant role the metaverse and AI can play in shaping future education.[18] One such platform embracing this potential is Whose Metaverse?, a learning platform that offers immersive courses designed to develop students' digital skills and foster collaboration.[19] Covering emerging technologies like metaverse basics, generative AI, and NFTs, the platform also emphasizes creativity. It uses a hybrid model, blending digital and physical learning spaces, where students gather at community hubs, such as one in Harlem in New York City, to collaborate and create using metaverse technologies. This approach is especially crucial for remote and marginalized communities, as a Stanford study has shown that immersive learning improves information retention by 75 percent.[20] Additionally, a study by PwC demonstrated a 275 percent increase in students' confidence, particularly in applying skills learned in immersive settings compared to traditional classrooms.[21]

I remember vividly walking across the Stanford University campus in the 2000s, attending seminars and classes, and hearing people discuss how what is now known as AR and VR would reshape the world. Stanford has long been a pioneer in immersive learning, exemplified by its Virtual People course, which has provided a fully immersive VR educational environment since 2003. This course allows students to explore applications in therapeutic medicine, sports training, engineering, behavioral science, and communication, showcasing the broad potential of VR and metaverse technologies.[22] My time on campus reinforced how forward-thinking institutions were already envisioning the future of immersive digital experiences—visions that are now becoming reality. Although innovation often takes time to gain widespread adoption, history shows that most major advancements can be traced back decades before they transform the mainstream.

While online learning has reshaped education, the metaverse promises to make current online learning methods seem rudimentary. AR and VR tools are game changers, offering safe, interactive, and immersive learning experiences that are more engaging and effective.[23] Virtual classrooms enable students to access education from anywhere in the world, opening new opportunities for those in remote areas or with travel difficulties. The metaverse can also deliver customized learning experiences tailored to individual needs and learning styles, allowing students to learn at their own pace and focus on areas where they need the most help.

AR and VR provide immersive experiences that go beyond traditional online learning management systems, making the sensation of being in the learning environment much more tangible. This heightened level of engagement enhances student participation and effectiveness. For example, Google Expeditions, a VR app, allows students to take virtual field trips worldwide, while Unimersiv, a metaverse platform, enables students to learn and collaborate in a virtual world. This depth of engagement in the metaverse far exceeds that in traditional online learning environments, making students more likely to pay attention and actively participate in learning activities.

What's Next?

COVID-19 rapidly accelerated the shift to online learning, forcing students and teachers to adapt to virtual classrooms, grading, communication, and engagement through videoconferencing and learning management systems (LMS). While these tools eliminated the need for physical classrooms, they fell short in delivering hands-on learning experiences—until now. The metaverse bridges these gaps by creating virtual classrooms, lecture halls, and laboratories, offering a more immersive and interactive learning environment.[24]

The metaverse also enhances educational interaction.[25] Unlike traditional asynchronous online environments, which can isolate students and instructors, the metaverse provides real-time social interaction that enriches the learning experience. Although the pandemic posed challenges for education, it also familiarized a broader audience with online learning and videoconferencing, setting the stage for more advanced virtual education.

During the pandemic, videoconferencing and other tools often struggled to meet the demands of effective pedagogy, particularly in K–12 education. While students were accustomed to video chats and online posts, these methods were not ideal for everyone, leading to issues like boredom, disconnection, and academic regression.[26] Education in the metaverse promises to merge the best of traditional teaching tools with AR/VR learning, breaking down barriers between the classroom and real-world experiences like field trips.

The pandemic has permanently altered teaching and learning, upskilling students and teachers in virtual settings, while businesses added remote training to the work-from-home model.[27] Immersive training has emerged as a powerful method for delivering soft skills and expertise. Across all educational levels and business training programs, the metaverse is offering tools that stimulate new levels of engagement. AR and VR are poised to play a pivotal role in driving future success. Leaders and organizations can best get involved by adopting and integrating small, proven technologies, building valuable experience in the process.

The metaverse is set to reinvigorate all levels of education and fields of learning. Its emerging tools can level the playing field, providing educational and economic access to historically excluded groups and communities. What was once difficult to teach online (like complex surgical techniques or massage therapy) becomes feasible in this new environment. Educators and leaders see exciting opportunities to revitalize learning and reintroduce personal elements lost in strictly online settings. Institutions like the University of Maryland, Baltimore County have created virtual campuses in the metaverse, where students attend classes and access resources like libraries. MIT's virtual laboratory allows students to conduct experiments in chemistry and biology, while UC Berkeley hosts lectures by Nobel laureates and offers virtual museum tours, bringing education to life in new and engaging ways.

The Learning Environment of the Metaverse

Educators and institutions are increasingly integrating AR and VR into web-hosted courses, profoundly enhancing the learning experience. AR-enhanced textbooks enable students to interact with 3D models, animations, and other dynamic content via their smartphones or tablets.[28] VR simulations provide safe, hands-on practice with complex tasks such as chemical experiments and surgical procedures. Additionally, interactive virtual tours and gamified learning experiences introduce new dimensions to education, boosting both engagement and retention.

The metaverse not only makes abstract concepts more tangible but also reintroduces the social interactions often missing in asynchronous learning environments. Real-time, immersive experiences in the metaverse bring back essential group dynamics—collaboration, camaraderie, laughter, emotional support, assistance, and immediate feedback—adding a human and kinesthetic element to the learning process. This environment seamlessly blends the best of traditional classroom pedagogy with the hands-on experience of a laboratory or internship. For example, an archaeology student could simultaneously engage in academic study and participate in a virtual dig at a historical site, effectively bridging abstract concepts with sensory experiences.

Studies underscore the benefits of such immersive learning. Research from the University of Southern California found that students using a VR platform for science learning performed better on exams and were more engaged than those who did not use the platform.[29] Similarly, a UC Irvine study revealed that students using a social VR platform for an English course participated more in discussions, collaborated more with peers, and felt a stronger connection to their classmates than those in traditional settings.[30]

The metaverse offers significant benefits in education, including increased engagement, enhanced collaboration, personalized learning, accurate simulations, and access to experts. However, challenges persist, such as the costs of setup and maintenance, access to necessary technologies, concerns about social safety, and the potential for addiction.

Impact on Colleges and Universities

Several institutions, including Fisk University and Florida A&M University, have embraced the concept of "metaversities," creating metaverse versions of their online campuses.[31] In partnership with VictoryXR, these institutions provide students with headsets and immersive experiences, setting a new standard for virtual education. Other companies are also entering this burgeoning market. For example, Classcraft engages students through gamification, quests, challenges, and rewards, while Nearpod uses AR and VR to create immersive learning experiences with simulations, games, and 3D models. EON Reality offers AR and VR resources for K–12 schools, universities, and businesses, covering subjects like anatomy, engineering, and history. Spatial develops interactive VR educational experiences for fields such as architecture, engineering, and design, serving customers worldwide. Additionally, Mozilla Hubs provides a free, open-source VR platform for collaboration and instruction, widely used by businesses and educational institutions globally.

Some experts predict that college education delivered through VR will become the norm. While the demand for educated, skilled graduates remains strong, many colleges and universities are experiencing declining enrollment.[32] To secure revenue, higher education providers have increasingly focused on nontraditional student populations, including adult learners, first-generation college students, students with disabilities, military personnel and veterans, immigrants, working students, and single parents. These learners, often constrained by responsibilities and schedules, frequently choose online learning, which can lead to feelings of isolation. The immersive environment of the metaverse could offer a valuable solution, providing these students with a more engaging and connected educational experience.

Transitioning to the Metaverse

Predicting the exact impact of the metaverse on kindergarten through PhD (K–PhD) education is challenging.[33] Schools face the dilemma of balancing the need to adopt new technologies with the risk of investing in what could be the next edtech fad. While leaders and policymakers in K–12 education recognize the inevitability of the metaverse, they may struggle to understand how it will fit into the educational landscape and whether it will promote equality or deepen disparities. Concerns arise about the future role of teachers—will they remain central, or will AI take their place? Educators and leaders are urging companies developing the metaverse to consider the unique dynamics of the educational environment as they design tools and resources, emphasizing the importance of creating safe, equitable, and inclusive spaces. VR learning holds the potential to deliver immersive and unforgettable academic experiences, bringing subjects to life in ways that leave lasting impressions and empower students to apply their knowledge confidently. However, this promise is tempered by concerns about inequality—Will all students have equal access to content and connectivity, or will a virtual divide emerge?

Despite these uncertainties, it is essential to begin exploring the metaverse. For teachers, this could mean integrating AR elements—already accessible via smartphone technology—into their teaching methods. While the subject matter remains the same, the learners' experience is transformed. Virtual colleges and metaversities may soon become commonplace. The metacurriculum won't change facts or events, but it will allow students to experience history, literature, or biology in ways that make them direct witnesses or participants. For example, a biology student might explore a cell's mitochondria, or law students could observe and even participate in famous trials. This approach promises to produce learners who are better prepared to engage with and solve real-world problems.

Ethical Issues in Metaverse Education

The rise of AR and VR in education introduces new avenues for academic dishonesty.[34] Students may exploit these technologies to access unauthorized materials during exams or collaborate unfairly on assignments. As these devices become more discreet, detecting cheating will become increasingly challenging. However, the same technologies that enable cheating can also serve as tools for detection in real time. AR and VR systems can monitor eye movements, facial expressions, and body language, offering insights into student engagement and emotional responses. Yet, this data collection raises significant concerns about privacy, security, and potential bias.

Colleges and universities are actively addressing these ethical challenges by developing policies, training faculty, and investing in technologies designed to prevent cheating in AR and VR environments.[35] Companies are also creating advanced plagiarism detection tools that leverage natural language processing, machine learning, and image recognition.

However, the ethical concerns are not limited to students. The vast amounts of data generated in metaverse-based education environments present both opportunities and risks. While analyzing this data can optimize learning and improve future educational tools, it also raises serious concerns about data exploitation and privacy. Protecting student information and ensuring transparent communication about data usage are paramount.

Beyond academic integrity, there are broader concerns about equity and seamlessness in metaverse education. Disparities in network quality, hardware access, expertise, and resources could exacerbate existing educational inequities. To fully realize the potential of immersive learning, educators must approach virtual education with a commitment to sustaining inclusion and equity, ensuring that all students have equal opportunities to benefit from these new technologies.

Health Care

Medical education will also undoubtedly benefit from metaversal technologies. A 2023 study by the Mayo Clinic found that VR-based pain management therapy effectively reduced pain and anxiety in chronic pain patients.[36] Similarly, surgeons at Johns Hopkins Hospital utilized VR headsets to visualize real-time images of a patient's brain during surgery, allowing them to avoid critical structures.[37] These examples, alongside AR-enhanced anatomy lessons and VR surgical training, signal transformative changes in medical education and health care delivery.

Health care is already stepping into the metaverse. Industry leaders anticipate that this immersive technology will revolutionize the field. A global survey of health care leadership revealed that 81 percent expect the metaverse to impact their organizations positively.[38] Many facilities and practitioners are eager to adopt 3D technologies, as the metaverse enhances the delivery and range of services that can be offered remotely. Patients have become accustomed to limited direct contact with health care providers, often relying on care delivered via telephone, videoconferencing, or AI. The metaverse, with its immersive, real-time experiences and sensory options, promises to enhance the provider–patient relationship further, offering new avenues for diagnosis and care. Health care leaders can expect AR and VR to expand possibilities and streamline service delivery in ways previously unimaginable.

Training Tomorrow's Health Care Professionals

Already, health care providers are learning to integrate the metaverse into their patient care. The Cleveland Clinic is utilizing VR to train residents in cardiothoracic surgery, allowing them to build skills and confidence by practicing complex procedures on virtual patients.[39] Similarly, the University of Central Florida College of Nursing's VR simulation lab enables nursing students to practice in various clinical settings, honing essential skills and knowledge.[40] At the Mayo Clinic, a mixed reality (MR) training program overlays 3D models of organs and other structures on real patients, enhancing students' understanding of human anatomy and the effects of diseases.[41] In the Netherlands, three hospitals—Amphia, MMC, and

Spaarne—have adopted VR to train their staff in high-risk procedures such as childbirth, surgery, and cancer treatment.[42]

The metaverse is poised to revolutionize how communities train and prepare health care professionals. Emerging AR and VR tools have the potential to significantly reduce the cost and complexity of teaching health care–related topics and training individuals in hands-on therapies and surgery. This capability allows for effective instruction across great distances, potentially reducing the shortage of qualified health care personnel in underserved areas worldwide.[43] Medical professionals could become truly borderless, able to serve patients and communities from any location.

The immersive environment of the metaverse promises to optimize training, particularly in skill development within the medical field. Simulated human bodies and facilities like emergency rooms and operating rooms can accelerate and refine the development of surgical skills. For instance, the University of Pittsburgh Medical Center is using VR to train surgeons in heart surgery, leveraging digital twins to enable precise skill refinement.[44]

This highly effective, accessible training offers the potential to close the personnel gaps that affect health care services and delivery in many regions globally. Communities with too few doctors or nurses could gain the hands-on training and certification needed to meet their health care needs.[45] These additional health care practitioners are likely to lead to improved health outcomes and lower mortality rates, which, in turn, could positively impact vulnerable, developing economies.

The Doctor Will See You Now

In 2020, the Cleveland Clinic began utilizing a metaverse platform to provide remote consultations for Parkinson's disease patients.[46] In 2023, the Mayo Clinic launched a virtual clinic specifically for cancer patients.[47] Health care providers are increasingly leveraging the metaverse to engage patients, educate them about their conditions, assist in making informed treatment decisions, and, when appropriate, connect them with others facing similar challenges. The health care sector is clearly ahead of the curve.

Remote diagnosis and patient monitoring have long been established practices, with telehealth and teletherapy now commonplace. The metaverse is expanding these possibilities even further. Metaverse telehealth services enable patients to consult with doctors from anywhere in the world, a practice the Mayo Clinic is employing to reach patients in rural areas.[48] These advancements have brought significant efficiency to the medical industry, increasing accessibility and bridging the gap between health care professionals and patients across distances.

Diagnosis and Treatment

The metaverse holds transformative potential in health care, offering remote diagnoses, pain management, and rehabilitation for patients recovering from injuries, strokes, or surgery. Immersive 3D environments allow patients to engage in exercises that restore motor skills and improve functional outcomes, providing health care professionals with insights beyond what traditional videoconferencing can offer. This shift from digital to virtual diagnostics promises faster, more accurate results, ultimately reducing costs.

For patients and caregivers, the metaverse enhances understanding of health conditions, supports the implementation of treatment plans, and improves interactions with medical staff. It also broadens access to medical knowledge through AI and other information sources.[49] Across health care services, treatments, and therapeutics, the metaverse's immersive capabilities elevate care experiences. Additionally, the integration of AR in communication resources enhances surgical procedures, advancing the quality and effectiveness of medical care.

Metaverse Mental Health Services

The metaverse is rapidly emerging as a powerful platform for mental health services, offering unique opportunities to create safe, supportive spaces where individuals can address their mental health concerns through avatars. Group therapy sessions can unite people from around the world, providing a shared space for those facing similar challenges, including victims and survivors of armed conflict. Various therapeutic approaches, such as individual therapy, exposure therapy, mindfulness, relaxation, and psychoeducation, can be seamlessly adapted to AR and VR environments. Companies like XRHealth or Your VR Therapy offer virtual treatment for anxiety, depression, and PTSD, while PatientsLikeMe uses the metaverse to deliver comprehensive mental health services.[50] Liminal VR and Floreo both develop experiences for individuals with autism spectrum disorder (ASD) to enhance social skills and reduce anxiety. Floreo also offers VR experiences that help people with dementia remain engaged and reduce agitation.

Mental health practitioners can customize virtual office settings to best meet their patients' needs, allowing for immersive 3D encounters that enrich the therapeutic experience. The US Department of Veterans Affairs is exploring VR as a tool for treating PTSD, enabling individuals to safely confront and process traumatic experiences in a controlled virtual environment. Beyond mental health applications, VR is also transforming life sciences, allowing researchers to visualize and manipulate complex biological structures in ways that were previously impossible. For example, scientists can now interact with protein-folding patterns in an immersive 3D space, improving their understanding of diseases and accelerating drug discovery.[51]

As the metaverse becomes a new platform for telemedicine and virtual health care, data storage and security will become critical. Blockchain technology offers a secure means to protect patient data, with decentralized servers enhancing security while still allowing physicians easy access to necessary information. What's more, existing patient care support apps, such as those that remind patients to take medications or schedule appointments, can be integrated with the metaverse, ensuring seamless communication and continuity of care across platforms.

Distant Healing

Health care in the metaverse heralds a new era of borderless medical practice, where regions with an abundance of providers can seamlessly support underserved areas.[52] The concept of a metaverse traveling nurse who can provide care without leaving home not only enhances service delivery but also offers social and emotional benefits for health care professionals and their families by reducing the need for extensive travel. The immersive 3D space of the metaverse re-personalizes the often-impersonal click- or call-only service experiences, a critical factor in effective healing and recovery.

While remote medicine can sometimes sacrifice personal interaction, the metaverse allows providers and patients to reconnect through avatars, maintaining the essential human element in care. Practitioners can also leverage bots to collect and process initial patient data, integrating these tools into broader diagnostic processes. The metaverse offers the potential for deep and accurate diagnoses, such as identifying eye conditions like glaucoma or neurological disorders by analyzing eye movements.

The metaverse's connectivity with mobile devices ensures reliable patient follow-up, and as more people engage with virtual health care, the adoption of these services will accelerate. Innovations in this area will focus on optimizing medical interactions and data handling, with speed and privacy as key priorities. Additionally, the development of virtual medical tools, such as scalpels and stethoscopes, is already underway.

The possibilities for health care in the metaverse are vast. Medical schools are incorporating VR and AR into their training programs, while patients are increasingly exploring the metaverse for health and wellness options. Nearly 90 percent of millennials and Gen Z are interested in care delivery and patient education in the metaverse, as well as its ability to connect them with distant family and friends during hospital stays.

The metaverse will enable practitioners to create targeted events and scenarios that benefit patients in novel ways. Alongside new technologies, existing smartphone and internet apps will continue to educate patients and caregivers and facilitate the monitoring of treatment plans. Leaders can anticipate significant and expanding opportunities to apply metaverse technology across the health care, education, and business sectors.

9 The Private Sector: Commerce, Banking, and Beyond

Around the world, businesses and global organizations are diving into the metaverse, transforming how they market, engage, and operate. Companies are creating immersive brand experiences, virtual showrooms, and interactive events, while financial institutions are exploring new ways to personalize customer service and integrate digital currencies. Real estate and construction industries are leveraging virtual tools to revolutionize property development and sales. The entertainment sector is hosting virtual concerts and events that draw millions, offering unprecedented access and interactivity. As the metaverse continues to evolve, these sectors are not just adapting—they are pioneering new strategies that blur the lines between physical and digital, setting the stage for a future where the metaverse becomes a central hub for commerce, real estate, entertainment, and beyond.

Marketing

Marketing in the metaverse is transforming the way brands interact with consumers, offering immersive experiences that go far beyond traditional advertising.[1] From virtual showrooms and product demos to gamified campaigns and influencer marketing, the metaverse allows brands to create dynamic, engaging environments that captivate audiences in new ways. In Nikeland, for example, users can create avatars, play games, and try on Nike products, while McDonald's hosts virtual scavenger hunts where participants can win prizes.

The metaverse's unique ability to blend fantasy with virtual reality makes it an ideal platform for innovative marketing and advertising. Brands are not only hosting or sponsoring virtual events, reaching global audiences, and building communities, but they are also enhancing consumer connections through virtual stores and worlds that offer immersive, interactive experiences. The use of NFTs is also gaining traction, with brands creating unique, tradeable digital assets that add a layer of exclusivity and value. Additionally, virtual billboards, branded concerts, fashion shows, sponsored content, and in-app advertising provide marketers with access to highly engaged audiences, making the metaverse a powerful tool for reaching and resonating with consumers.

Marketing 3.0

Marketing 3.0 emphasizes creating human-centric experiences, where consumers seek not just products and services but meaningful and engaging interactions. This approach is perfectly aligned with the metaverse, where brands can establish virtual worlds that offer consumers personal, immersive experiences with their products. Companies can craft virtual events designed to both entertain and inform, connecting with consumers on a deeper level than ever before.[2]

Marketing played a pivotal role in shaping the internet, and it is poised to do the same for the metaverse. While this new frontier offers exhilarating opportunities, it also presents challenges for marketers who may initially feel intimidated by the need to adapt. However, the potential to allow consumers to test drive, try on, and fully experience products in a virtual environment—along with the social warmth of interacting with salespeople, even if they are bots—makes the effort worthwhile. Leaders should feel encouraged to explore this new territory, as early adopters are likely to see significant rewards.[3]

Tech giants like Apple, Google, and Microsoft, along with smaller companies, are all staking their claims in the metaverse, recognizing it as a powerful platform for marketing and advertising. Wherever people gather, marketing follows, and the metaverse already supports a range of activities—work, collaboration, communication, and socializing—that naturally attracts marketing efforts. As businesses of all sizes turn to the metaverse, it is becoming the next big opportunity for marketing and sales.

The metaverse is a fresh, new frontier, with the underlying structures—hardware, software, speed, and content—now viable thanks to advancements in Web 3.0. Users can immerse themselves in digital versions of their favorite real-world activities, from dining to roller skating, simply by donning a headset and haptic gloves.

Investment in augmented reality (AR) and virtual reality (VR) has surged globally, signaling that the metaverse will soon represent a significant source of value.[4] From $12 billion in 2020, investment was projected to reach over $70 billion by 2024.[5] Businesses are pouring resources into figuring out how to harness the potential of this new landscape, making marketing in the metaverse not just feasible but highly profitable. As 2025 progresses, the focus is shifting from experimentation to scaling, as early investments begin to deliver measurable returns.

The Best of Both Worlds

The metaverse offers a unique opportunity for marketers and businesses to blend and bend the boundaries between virtual and real worlds. In these hybrid spaces, where the line between the metaverse and real life becomes blurred, businesses face the challenge of managing perceptions and brand identities in a way that is both consistent and advantageous. As virtual worlds increasingly overlap with real-life activities, new consumer behaviors are bound to emerge. Marketers must be vigilant in recognizing these trends and engaging with them in ways that support both customer well-being and the aspirations of their companies.

Consistency in messaging is crucial to maintaining the essential balance between a business's real-world identity and its presence in the metaverse.[6] While individuals may explore alternate identities in digital realms, businesses cannot afford to be inconsistent. A company's metaverse persona must remain closely aligned with its real-world brand to reinforce consumer trust and familiarity. This sustained link helps transfer the positive sentiments consumers associate with a brand in real life to their experiences in the metaverse.

As new users navigate the virtual world, they may experience uncertainty. Familiar brands, logos, and products serve as anchors, providing a sense of psychological safety.[7] In this context, marketers have a significant opportunity: they can either strengthen or disrupt consumer relationships. While the metaverse offers the allure of breaking away from traditional patterns, companies should consider the value of consistency across both real and virtual worlds. By maintaining clear and aligned messaging, businesses can create a seamless experience that benefits both consumers and marketers, reinforcing trust and brand loyalty.

New Feats of Marketing

The immersive 3D environments of the metaverse present a groundbreaking opportunity for marketers to pioneer novel advertising strategies. Those who have yet to dive into AR should stop hesitating. It's better to try and fall short now because the experience and insights gained will provide a competitive edge over those who enter the space later, even with improved technology. Consumers are already engaging with the metaverse, showing far less trepidation than many businesses or their leaders.

The metaverse is advancing rapidly, and now is the time for marketers to move in. The current phase of development won't last forever. Once a critical mass is achieved, the metaverse will be fully operational, similar to the early days of the internet. As the infrastructure continues to grow and user adoption increases, the metaverse will become an integral part of the digital landscape. Marketers, leaders, and organizations have a unique window of opportunity to learn and innovate by getting involved early.

Future metaverse marketing technologies are likely still in development, or even yet to be invented.[8] In the meantime, marketers should experiment and capitalize on the current wave of excitement among consumers and investors. There's an air of anticipation—something huge is on the horizon, and it's almost ready. Companies like Epic Games, which raised more than $2 billion to develop its metaverse presence and partnered with LEGO to create a kids' metaverse, demonstrate that now is the time to adopt a forward-thinking mindset.[9] By experimenting today, marketers can refine their strategies and build brands that will thrive in the metaverse.[10]

The coronavirus pandemic, with all its challenges, gave a significant boost to the metaverse. The virtual space offered users a safe, engaging environment where they could create and interact as unique identities, free from the constraints of the real world. This sense of agency and interaction likely helped users bond with the medium, as they found solace and connection in an otherwise disrupted reality. The gaming industry, which saw the potential early on, built responsive platforms that have paved the way for other businesses. Now, the metaverse is a vital space for brands wanting to keep millennials and Gen Z engaged and aware of their products and services.

Metaverse platforms like Roblox and *Fortnite* offer extensive opportunities for brands. For example, Vans created a virtual skate park where users could try its products, identifying the metaverse as the ideal place to reach its core demographic of 13-to-35-year-olds.[11] Similarly, Gucci's 2021 launch of the Gucci Garden on Roblox allowed visitors to mingle and purchase digital items, showcasing the potential for luxury brands in this new space.[12]

As the metaverse expands beyond gaming, other sectors are finding opportunities to market their products and services. Virtual conventions and trade shows, like the TECHSPO Technology Expo and SIGGRAPH Conference, allow attendees to engage from anywhere in the world.[13] Car shopping and shows have also embraced the metaverse, offering immersive brand experiences. In this environment, marketers have the chance to run cost-effective campaigns, experimenting with different approaches to find what works. The old methods of digital advertising may soon look as outdated as early internet marketing, where a scanned flyer was considered innovative. Optimizing ads for the metaverse is crucial, and those who enter early will set the standards for future marketers. Engagement remains the key metric for success, offering invaluable feedback for those willing to take the plunge.

Get In!

As companies pivot to embrace the future, just as they once appointed chief information officers, they are now installing chief metaverse officers. A quick LinkedIn search reveals the increasing prominence of the metaverse in headlines, professional titles, and work experience. The emergence of roles like chief of metaverse marketing is just around the corner. Brands such as Disney, Gucci, Nike, and Facebook are already pioneering the creation of virtual settings, events, currencies, art, apparel, and assets, effectively establishing a new economy and a fresh marketing dynamic.

Globally, consumers aged 13 to 34—the prime demographic for the metaverse—hold a staggering $4.4 trillion in spending power.[14] Mastering the art of understanding and engaging this group will be crucial for future metaverse marketers. In this evolving landscape, consumers are the driving force behind successful marketing strategies, and the metaverse will, in turn, reshape consumer behavior. Navigating this space to maintain brand awareness and maximize earnings will require a deep and early engagement with the still-developing metaverse.

Buyer, Beware . . . Much More Aware

As the digital universe expands, so does the potential for fraud, spreading its reach across industries and exploiting vulnerabilities in the as-yet unregulated metaverse. This emerging virtual world has quickly developed a dark side, mirroring the deceptive tactics found in the physical world.[15] For instance, the finance worker who was tricked into paying out $25 million after a video call with a deepfake "chief financial officer," and the infamous "rug pulls" in the crypto world, illustrate how easily developers can disappear, leaving investors with nothing.[16] Similarly, pump-and-dump schemes manipulate virtual asset prices, preying on the uninformed in a market where assets like virtual land and avatars are particularly vulnerable to exploitation.

This issue isn't confined to a single sector within the metaverse but affects its entire spectrum. Scams like fake metaverse projects underscore the urgent need for standardized marketing and advertising practices. Beyond the high cost of entry, the lack of established guidelines increases the risk of fraud for consumers. Enticed by the allure of cutting-edge virtual assets, consumers often fall victim to counterfeit NFTs and digital real estate—perfect digital replicas that hold no value. The metaverse, rife with the risks of sophisticated phishing pump-and-dump schemes, and other scams, demands the same vigilance we've learned to apply in traditional digital spaces: diligent research, careful clicking, strong passwords, two-factor authentication, and a healthy skepticism toward deals that seem too good to be true.

While marketing in the metaverse may be more technology-infused and exciting, sales might evolve beyond the current internet's practices. A sales avatar, whether human or bot, interacting with a customer can gather and retain information, potentially rendering traditional cookies obsolete. This richer customer dataset, collected during and after immersive, sensory-enhanced interactions, could redefine marketing insights. The metaverse reintroduces the salesperson in a more dynamic role. Immersive marketing and hands-on sales experiences provide businesses with multiple opportunities to learn about customers and refine their strategies. Businesses anticipate that items or services sampled and purchased in the metaverse will have lower return rates. Leaders should expect immersive 3D marketing to become a game-changing tool, but they should also prepare for the

The Private Sector

availability of user-friendly options that don't impose a steep learning curve on consumers.

Consumer Sector and Metaverse Commerce

Businesses entering the metaverse will find that consumers already have interactive AR and VR options. H&M and Target are leading the way in the metaverse by allowing virtual shoppers to try on clothing, while Sephora offers a similar experience for makeup.[17] Tiffany & Co. has introduced a virtual tool that lets users overlay jewelry onto their hands or necks, enhancing the shopping experience.[18] Warby Parker's AR try-on tool allows consumers to see how a pair of glasses will look on them before making a purchase, bridging the gap between virtual and physical retail.[19] Walmart is pushing the boundaries with a virtual store where customers can shop for groceries and other items using VR technology and have them delivered directly to their homes.[20] Similarly, Kroger's pickup assistant enables voice-activated grocery shopping, and Kohl's offers a virtual stylist that recommends flattering clothes and accessories for purchase.[21] CarMax and Toyota provide virtual showrooms where customers can explore different vehicle models, make purchases, and arrange deliveries.[22]

Buying and selling are not disappearing anytime soon, even as new technologies emerge. New tech typically doesn't displace thriving markets overnight. The key takeaway for businesses is to continue selling what they have been but to use the metaverse to make marketing and purchasing more exciting than they have been in years. At the intersection of marketing, sales, goods, and services, the metaverse offers a fresh approach to familiar activities. Consumers are still purchasing the same types of items; they're just shopping in a new AR/VR space. The tangible products they buy still arrive in boxes and packages, but the journey to purchase has become an immersive experience, transforming normal e-commerce into something far more engaging and interactive.

What's for Sale?

In addition to tangible goods, consumers in the metaverse are increasingly purchasing intangible items such as NFTs, virtual apparel, and even virtual vacations.[23] They are spending real money on digital experiences, from virtual tours to immersive travel adventures. Outfitting avatars with digital apparel and accessories, often incorporating NFTs, has become a significant aspect of this virtual economy.

The term "metaverse" often brings to mind the advanced technology that drives it—the existing tech and the groundbreaking innovations necessary to create an ideal virtual world. While innovation is crucial to fully realizing Web 3.0 and enabling rich, immersive experiences, it's easy to become overly focused on the technology and overlook the people on the other end of the devices and connections. Interestingly, over 70 percent of consumers report that they plan to use the metaverse for purposes beyond gaming within the next two to five years.[24] They express this intent regardless of the current state of technology, indicating a readiness to embrace whatever is available. This trend suggests that businesses and markets should prioritize understanding consumer desires and expectations in the metaverse, focusing on delivering experiences that align with what users want to do or experience in this emerging digital space.

Who Wants to Jump In?

An Accenture survey reveals that 55 percent of consumers are interested in becoming regular users of the metaverse, with only 4 percent focused exclusively on gaming.[25] This indicates a broader, more mainstream perception of the metaverse, where consumers anticipate using it for everyday tasks like bill payments, health care, and learning. To succeed in this evolving landscape, businesses must understand and align with consumer needs and expectations, anticipating and exceeding the experiences that users are looking for.

Consumers have clearly defined areas of interest in the metaverse. According to Accenture, demand is emerging in five primary sectors: 36 percent of consumers are eager to explore virtual travel, 37 percent seek access to medical care, 40 percent are interested in retail and e-commerce, and 41 percent are drawn to recreation and exercise.[26] For these users, the metaverse is both personal and practical—a space designed to enhance daily life rather than serve purely as an escape.

They prioritize ease of use and functionality over flashy, sci-fi-inspired experiences. They seek a metaverse that works—one that delivers practical solutions and facilitates everyday interactions with family, friends, and services. This pragmatic attitude is guiding businesses, designers, and engineers to focus on creating a metaverse that meets real-world needs.

As consumers continue to embrace the metaverse, they are concentrating their time and spending in specific areas. They want to attend appointments, shop, and connect with distant loved ones in immersive, interactive ways, all without leaving home. These consumer priorities provide businesses with a clear roadmap for success. Understanding and responding to these preferences ensures that companies deliver what consumers truly want. For example, in the realm of health and fitness, consumers have long turned to Web 2.0 and social media for at-home exercise content. In the metaverse, they seek even more privacy, affordability, and ease of use, rather than perfection in hardware or software.

Businesses must recognize that the technological landscape of the metaverse is constantly evolving.[27] While new developments will inevitably make current hardware obsolete, the key to long-term success lies in staying focused on the consumer. By hearing and responding to consumer needs, businesses can ensure that every iteration of the metaverse meets expectations and fulfills consumer demands. In the near term, moving in step with the consumer is just as crucial as engineering and design, ensuring that the metaverse remains relevant and valuable to its users.

Consumer-Centric

The strategy for success in the metaverse hinges on a continuous understanding and learning about consumers. Businesses and consumers must embark on this journey together, with businesses shifting their perspective from viewing consumers as mere targets to recognizing them as partners and co-innovators. This strategic mindset must be sustained into the future, where organizations that prioritize consumer feedback as a central element in design and innovation will thrive in the metaverse.

Consumers enter the metaverse with their existing brand awareness and loyalties, expecting the immersive media to offer richer and more complete experiences. The metaverse sales experience must surpass what is currently available on the internet, providing a more engaging and satisfying interaction. It has the potential not only to exceed the quality of online shopping but also to rival and even surpass the in-store experience. In the metaverse, shoppers can feel as though they've left home, visited a store, and been personally attended to, all within an emotionally dynamic environment. Importantly, businesses have significant control over shaping these emotional experiences, enhancing consumer satisfaction and brand loyalty.

Unlike the current Web 2.0 shopping experience, which consumers primarily use for convenience, the metaverse promises an interactive, immersive setting that consumers are eager to explore. Brands will need to carefully translate their signature in-store dynamics to the virtual world, maintaining familiarity while meeting and exceeding consumer expectations. This approach will create a smoother transition for marketers and salespeople, making the metaverse an extension of the brand rather than a completely new environment.

In the metaverse, the relevance of cookies diminishes; customers are no longer just clicks to be tracked. Instead, the focus shifts to conversation and service. The metaverse sales experience, with its rich data yield, offers insights that remove much of the guesswork from personalized marketing. The immersive, one-on-one nature of interactions captures the customer's full attention, leading to more satisfying purchases and stronger brand loyalty. The metaverse represents a new frontier where businesses can deepen their relationships with consumers by creating experiences that are as emotionally engaging as they are technologically advanced.

How It's Going

Businesses can capture sales opportunities in the emerging metaverse by expanding their use of immersive technologies and exploring innovative marketing strategies. Following IKEA's lead with its AR Place app, companies can create similar experiences that allow customers to visualize products in their real-world environments. For example, furniture brands could let customers virtually rearrange entire rooms, while home improvement stores could offer tools to simulate renovations.

Another promising approach is *business-to-avatar* (B-2-A) marketing, where businesses cater to the needs of avatars within the metaverse.[28] Just as toy companies market to children while recognizing parents as the ultimate purchasers, brands can position their products and services for avatars, knowing that the decisions are made by real consumers. Avatars, being entirely virtual, still have needs that businesses can fulfill, such as clothing, accessories, and virtual real estate, making B-2-A a significant new channel for commerce.

Additionally, businesses should consider creating NFTs (non-fungible tokens), which can be used to offer certified, one-of-a-kind digital items. NFTs allow companies to designate custom art, apparel, media, and other unique content, providing value and helping to build brand awareness. By offering exclusive digital assets, brands can tap into the growing demand for collectibles and personalized virtual experiences.

Exploring less mainstream selling platforms is another way to capture sales opportunities in the metaverse. Platforms like Microsoft Mesh, Horizon, Twitch, DressX, and Clubhouse offer unique environments where businesses can connect with new audiences and experiment with different types of virtual commerce. These platforms may not yet have the same reach as more established ones, but they present opportunities to engage early adopters and innovators, setting the stage for future growth.

Leaders should recognize that the bulk of revenue expected from the metaverse will come from the traditional business of buying and selling consumer goods, now enhanced by immersive technologies. While the metaverse may seem like uncharted territory, it is essential for companies to form a strategy and get involved in metaverse commerce. There's ample room to test and experiment while minimizing risk, and early involvement will provide businesses with a competitive advantage over those who wait for the metaverse to mature.

Getting in early, even with modest returns, will allow companies to become familiar with the dynamics of this new consumer sector. Over time, these experiences will yield insights that can be leveraged as the metaverse continues to evolve, positioning businesses to thrive in this exciting new marketplace.

Financial and Professional Services

A sector to which the metaverse is poised to bring a much-needed infusion of personalization is the banking and financial services industry. Historically, banks have been slow to fully embrace new digital trends, as evidenced by their initial reluctance to adopt the internet. For example, the UK's Nottingham Building Society was a pioneer in offering online banking services as early as 1983 with its Homelink system, however, US banks like

Wells Fargo waited until 1995 to introduce online banking. Fortunately, the lesson has been learned, and banks today are not waiting to dive into the metaverse.

Early in 2022, JPMorgan Chase became the first bank to stake its claim in the metaverse by establishing the Onyx Lounge in Decentraland, a virtual marketplace for digital currencies located in Metajuku—a metaverse version of Tokyo's Harajuku district.[29] Following JPMorgan Chase's lead, other financial institutions like KB Bank in South Korea, HSBC, and American Express have made moves to enter the metaverse, exploring virtual branches, purchasing virtual land, and filing trademark applications related to NFTs.

The metaverse offers significant advantages for both customer and financial services, presenting an opportunity for banks to reintegrate the social component that was lost when banking moved online. Two-thirds of industry decision-makers believe the metaverse will positively impact the banking industry, and more than a third anticipate a breakthrough.[30] The introduction of cryptocurrencies and NFTs—native assets of the metaverse—presents banks with new challenges, but also with fresh opportunities to remain competitive and relevant.

As financial institutions navigate these new digital landscapes, the metaverse promises to broaden their horizons, offering a new array of assets, technologies, and experiences that can enhance customer engagement and drive innovation in the industry.

The Phases of Banking

Banking has evolved significantly over the past 30 years, transitioning through distinct phases that have reshaped the industry. From the traditional savings and loan systems of the 1800s, characterized by central banks and physical branches, the industry has embraced digitization over the last decade or two. This transformation enabled banks to offer new digital services and shifted the focus more toward customer-centric operations. The advent of internet banking marked a significant change, allowing banks to interact with customers in more direct and accessible ways. More recently, the move toward open banking has seen financial institutions sharing customer data with third-party developers, fostering the development of innovative products and services.

In recent years, blockchain technology has revolutionized the sector by enabling the creation of digital currencies and NFTs. This technology has also helped banks reduce risk and fraud, thereby enhancing security. Now, the metaverse is pushing the boundaries of banking even further, offering the potential to reduce operating costs while reaching and serving new markets in a more personalized and customer-centric manner.

Through AR and VR, banks can offer a warmer, more immersive experience. Customers will be able to open accounts, apply for loans, and receive personalized financial advice without ever visiting a physical branch. This shift will not only lower operating costs for banks with physical locations but also create more engaging and personal experiences for online-only banks. By leveraging gaming expertise, banks can incorporate elements from the gaming industry—such as interactive tutorials, gamified financial planning tools, and immersive customer service avatars—to make banking more engaging and user-friendly. Features like real-time financial simulations, achievement-based rewards, and interactive dashboards can help customers better understand financial concepts, track their spending habits, and make more informed decisions. These techniques, commonly used in gaming to enhance engagement and retention, can increase customer satisfaction and loyalty by making banking feel more intuitive and enjoyable.

Moreover, the metaverse provides an opportunity to enhance employee experiences, making banking a more dynamic and interactive profession. Virtual reality training programs, collaborative workspaces, and artificial intelligence (AI)-driven simulations can improve staff development, efficiency, and engagement, fostering a more adaptable and tech-savvy workforce.

The potential of the metaverse extends beyond just banking. Financial advisers, for instance, can utilize AR and VR environments for data modeling and simulation, making it easier to explain complex financial concepts like risk to clients. These immersive tools allow for more effective communication and decision-making, enhancing the overall client experience.

However, the current state of financial services in the metaverse is still experimental. Companies venturing into Web 3.0 are navigating uncharted territory, recognizing the need to engage with emerging technologies to become future leaders in a mature metaverse. The pandemic accelerated the adoption of the metaverse as a virtual escape, and forward-thinking organizations like Delaware-based bank ZELF have already begun to innovate in this space by introducing unconventional lending practices that allow customers to convert digital collateral into real-world money, signaling the beginning of hybrid and innovative practices that will define banking in the metaverse.

This experimental phase is crucial as it sets the stage for the future of financial services, where early adopters will likely emerge as industry leaders when the metaverse reaches full maturity.

Doing New Things with the Same Old Money

Over the past three decades, banking has been reshaped by technological advancements, with blockchain technology recently contributing critical enhancements to security. NFTs and other tokens are providing banks with new opportunities and diversified areas of interest and operation. As the metaverse continues to develop, some banks are already devising policies to manage the movement of monetary instruments within this virtual realm.

The emergence of banking in the metaverse is expected to reshape the industry, much like consumer-centered commerce. Banks are poised to benefit from the re-personalization of interactions with customers, a dynamic that is also anticipated to transform the financial services sector. The introduction of human interaction through digital means allows for better adaptation of advisory and informational services within the metaverse. The immersive, interactive nature of the metaverse enables advisers and clients to gain accurate insights into each other's needs, fostering trust and credibility, which are essential in financial services. The metaverse offers professionals new avenues to invite, meet, and engage with existing and potential clients.

Beyond banking, the metaverse is integrating related industries such as investment and trading, legal services, insurance, and consulting. Investment firms can build virtual trading floors where investors track markets and make trades in real time. Insurance companies can create virtual simulations of accidents to help customers understand their coverage. Lawyers and consultants can conduct meetings and run simulations with clients in these virtual environments.

At the heart of what the metaverse brings to banking and financial services is the concept of re-personalization. The metaverse is poised to reinfuse banking and financial services with social dynamics that have been diminished in the digital age. Some banks are already assigning avatars to employees and facilitating services on various metaverse platforms, with loan officers, financial planners, advisers, and other professionals eager to engage clients in these immersive environments.[31] The presentation of information, client education, and the overall customer experience are expected to benefit significantly from the sensory-rich elements reintroduced by the metaverse. Leaders in the industry can look forward to an array of new possibilities as the metaverse enhances the human aspects of banking and professional services.

Travel and Tourism

In another sector, companies are increasingly exploring and creating experiences tailored for travelers and tourism within the metaverse. As this virtual world expands, it is expected to eventually offer virtual equivalents of everything in the real universe, making virtual travel a natural and integral component of metaverse activity and economics.[32] During the pandemic, museums worldwide provided free, limited virtual access to their collections, and many well-known sites introduced live cams, allowing people to visit these locations remotely. This concept of virtual tourism can serve as either an end in itself or as a way to sample and promote real-world vacation destinations.

The metaverse is revolutionizing the role of digital technology and the internet in the travel and tourism industry, transforming it from a tool for sampling or simulation into a platform capable of delivering the experience itself. Many websites have long aimed to immerse users in rich audio and visual experiences, while various devices and smartphone apps strive to create the sensation of being elsewhere. The metaverse elevates these efforts, offering leaders in the travel industry elaborate options and opportunities that align with diverse business capabilities and goals.

Whether as a means of enhancing real-world tourism or as a standalone virtual experience, the metaverse is poised to significantly influence the travel industry. The immersive digital traveler already has options within the metaverse, and future developments will likely continue to blur the lines between virtual and real-world tourism.

Here to There from a Headset

Virtual tourism presents a range of benefits for travelers, particularly in making inaccessible, dangerous, or prohibitively expensive destinations more reachable.[33] The metaverse allows virtual visitors to experience these settings in a safe and controlled environment, with the added advantage of immersive elements that enhance intercultural learning and connection. Both VR and AR play crucial roles in this space. For instance, services and guides in the metaverse, possibly just an AI bot away, can help visitors navigate unfamiliar cities, offering real-time assistance and cultural insights.

Amazon Sumerian already pioneered AR apps that overlay information about landmarks and attractions, enriching the virtual travel experience. Similarly, Matterport offers virtual tours of iconic sites like the Eiffel Tower and the Great Wall of China, providing an immersive experience that rivals physical visits. As this technology continues to evolve, a VR headset might soon become as essential for travel as a passport, redefining the way we explore the world.

Choose Your Destination

The metaverse is uniquely positioned to transform the way people experience vacations, especially those centered around visits to famous sites or museums.[34] Many individuals have bucket lists filled with the remnants of ancient civilizations, such as the Roman Colosseum, the Egyptian and Mexican pyramids, and the Great Wall of China. Iconic landmarks like the Taj Mahal, Niagara Falls, the Eiffel Tower, Big Ben, and the Statue of Liberty are also top destinations. Among these, the Great Pyramids of Giza, Niagara Falls, and the Great Wall of China stand out as the most visited sites globally, attracting over 10 million visitors annually.[35]

Virtual reality tourism offers the potential to significantly increase the number of people who can "visit" and experience these sites while also helping to preserve them by reducing physical wear and tear. Through VR, people can explore these wonders without ever leaving home, accessing a level of immersion that rivals the real-world experience.

Adventure and spectacle are also central to many vacations, with people traveling long distances for activities like white water rafting or safaris. The metaverse, through AR and VR, makes these adventures accessible without the associated costs or health and safety concerns. With just a headset and a pair of gloves, users can experience faraway destinations as if they were really there.

The same technology that brought major artists like Travis Scott and Ariana Grande to *Fortnite*, or the Foo Fighters to Meta, can similarly bring the masterpieces of DaVinci, such as the *Mona Lisa*, right to you. In this new era of virtual tourism, avatars will have the freedom to explore tombs, temples, ruins, and famous landmarks, offering an unprecedented level of interaction with the world's most treasured sites. Whether it's encountering priceless works of art in museums or exploring historic sites, the metaverse promises to redefine how we think about and experience travel.

The New Wave of Tourism

Like many aspects of the metaverse, virtual travel and tourism represent a two-way street. What happens in the metaverse won't stay there—it influences and enhances real-world experiences. A visit to a city or attraction in the metaverse can serve as a satisfying experience on its own or as a preview, sparking the desire for an actual visit. The metaverse can even assist in sampling, selecting, and booking accommodations for real-world trips. Conversely, visitors to a real-world site might follow up with a VR or AR revisit, allowing them to focus on and explore particular aspects they enjoyed during their physical visit.

Virtual tourism also acts as a powerful marketing tool for in-person experiences.[36] For instance, Seoul's presence in the metaverse functions this way, giving virtual tourists a taste of the city's sights and sounds, potentially inspiring a real-life visit. Similarly, Madrid has launched an immersive initiative showcasing around 40 of its top attractions, helping potential visitors make informed decisions. This approach allows the metaverse to sample real-world experiences in ways that traditional marketing never could, enhancing and motivating the desire for real-life travel.

For these experiences to be successful, however, the technology driving the metaverse must perform consistently and optimally. Network latency and outages can be particularly disruptive in immersive environments, breaking the illusion and potentially ruining the experience. As AR and VR headsets become smarter and lighter, and networks reduce latency while improving graphics rendering, seamless metaverse travel and tourism become more attainable.

People who have experienced the same site both in real life and in the metaverse report that each experience enhances the other. A metaverse visit before a physical trip can help visitors appreciate the details and plan their time better, while a follow-up virtual visit allows them to explore and revisit highlights they may have missed.

The coronavirus pandemic dealt a heavy blow to global tourism, but the metaverse offers two streams of hope: a fully immersive way to virtually travel and vacation, and a new marketing avenue for attracting tourists to real-world destinations. The future of tourism now intersects with blockchain, free from time constraints, high costs, and the hassle of airports.

The metaverse also offers the possibility of combining experiences in unconventional ways. Imagine cruises to real or fantastical destinations themed around specific games, cultures, historical periods, or academic interests. Business travel, conventions, and trade shows are also reimagined in VR, with companies like Disney and other entertainment groups likely to lead the way. Beyond virtual tourist dollars, businesses have a unique opportunity—and obligation—to protect and promote their brands in this new frontier.

The Many Benefits of Travel from a Headset

Travel in the metaverse holds immense promise, potentially triggering beneficial changes across the tourism industry. The speed and reduced costs associated with virtual travel mean that consumers can explore more destinations in less time and with greater frequency. As Mark Twain famously noted, "Travel is fatal to prejudice, bigotry, and narrow-mindedness," and immersive travel experiences facilitated by the metaverse could significantly contribute to fostering intercultural understanding.[37] As the virtual travel industry begins to take off, governments and the public sector will play a crucial role in duplicating and maintaining many tourist attractions, ensuring their availability to a global audience.

Mainstreaming virtual travel can create well-informed global citizens while simultaneously relieving the physical stress on heavily visited sites. It offers the potential to preserve iconic locations long after they are damaged or lost due to natural disasters or human activities. Virtual tourism enables visits to famous sites before they were affected, and by opting for metaverse travel, tourists can help reduce CO2 emissions from fuel-intensive modes of transportation.

The metaverse and its underlying technology are actively shaping the future of travel. Industry experts are already describing the emerging opportunities as the next big thing in the sector. A growing number of leaders in travel and leisure are warming up to these possibilities. According to Accenture's Technology Vision 2022 global business and IT executive survey, 71 percent of industry executives believe the metaverse will positively impact their businesses, with 42 percent predicting a significant shift.[38] If early engagement with this technology is any indication, the growth of virtual travel may be rapid.

Travel in the metaverse feels natural and intuitive, especially as the transition from gaming and entertainment to virtual visits and tours is smooth. The hard-hit tourism industry can leverage the metaverse to generate new revenue streams. Businesses can use the metaverse to drive real-world activities, with visitors to virtual sites competing to win real-world prizes. Additionally, the metaverse can serve as a powerful marketing tool, introducing and selling in-person travel packages. Potential real-life travelers can use the metaverse to sample accommodations, hotels, and attractions, or passengers on physical flights can participate in fully virtual games and experiences, all this allowing for immersive previews and seamless purchases in one virtual space.

Beyond supporting existing tourism activities, the metaverse offers tools to enhance operational efficiency within the airline and travel industry.[39] Digital twinning technology can help airlines plan and schedule more effectively, perform maintenance, and improve safety.

One of the key benefits of entirely virtual tourism is its potential to reduce carbon emissions. While travel for business may become nearly obsolete in the metaverse, leisure travel—whether for vacations, family gatherings, or events—may not initially see the same reduction. However, the overall decrease in fuel demand, scaled-back operations at travel hubs, and other related functions offer considerable promise. Even though the metaverse itself consumes significant energy (which we will discuss in the next chapter), it also presents opportunities to save fuel and reduce carbon footprints.

For leaders in the travel and tourism industry, the message is clear: Engage with the metaverse now, even if on a limited scale. Practicing and testing the possibilities of the metaverse will prepare enterprises to scale up without significant challenges. Those who delay may find themselves unprepared for the new realities of virtual tourism. The metaverse is here to stay, and early adoption will be key to thriving in this rapidly evolving landscape.

Real Estate, Construction, and Planning

The next topic to examine shifts us away from tourism and back toward financial activities. The real estate and construction industries are not hesitant to invest in emerging technologies, signaling their belief that the metaverse is not just the next big thing but is already here and operational for various industry applications.[40] Virtual property tours, immersive real estate listings, and virtual auctions have already become part of the industry standard, demonstrating the readiness of real estate to fully embrace the metaverse. These sectors are particularly eager to harness the power of digital twins, which offer simulation tools for complex planning, design, collaboration, evaluation, and problem-solving. Whether it's a small addition, a tiny dwelling, or large-scale building projects, AR and VR technologies allow for planning and revisions without the costly mistakes or failures traditionally associated with construction.

Real estate and construction leaders must be keenly aware of the vast applications of AR and VR technologies and how these tools can enhance success and drive earnings. Companies like Matterport are leading the charge, offering potential buyers a more immersive experience by building 3D models of real estate properties that can be viewed in the metaverse. Similarly, PropTech360 provides VR property tours, allowing potential buyers to explore properties immersively from any location. Another company, REX, enables real estate agents to create virtual listings, offering buyers the ability to visit and purchase properties without ever leaving their own homes.

These advancements not only streamline the real estate buying process but also broaden access, allowing prospective buyers to explore properties globally with ease. As these technologies continue to evolve, they will likely become an integral part of the real estate and construction industries, further solidifying the metaverse as a vital platform for the future of property development and sales.

Changing Times, Changing Ways

Another way that the metaverse is revolutionizing the construction and real estate development industries is by offering new ways to approach design, planning, and collaboration. Virtual design and construction (VDC) is now being used to create detailed virtual models of buildings and infrastructure. These models allow for the simulation of construction projects, helping to identify and address potential issues before they arise. Leading architecture and engineering firms like HOK and AECOM are already utilizing the metaverse in this capacity. Additionally, companies like Siemens, Microsoft, and Trimble are developing metaverse software specifically designed for construction and urban planning.

These advancements are creating unexpected opportunities for architects, builders, and a broad range of allied professionals. As the metaverse becomes more integrated into construction and real estate, it highlights the interconnectedness of various elements within the industry. This shift is not only influencing the physical world of real estate but also shaping the virtual real estate markets within the metaverse. Three key areas of construction and real estate that stand to benefit directly from the metaverse are construction, interior design, and sales.

The metaverse facilitates real-time collaboration among different stakeholders in a construction project, such as architects, engineers, contractors, and owners, regardless of their physical location. This capability enhances the efficiency and effectiveness of project management. Furthermore, the metaverse can be leveraged to involve the public in neighborhood and urban/regional planning by creating virtual models of proposed projects and inviting community feedback.

The early adoption of the metaverse in construction and real estate exemplifies a dynamic trend seen across various sectors: a continuous flow between real life and the virtual worlds of augmented reality and virtual reality. This bidirectional influence fosters innovation and new possibilities. By using VR to plan, pilot, and optimize construction methods, buildings, neighborhoods, and municipalities, the industry can achieve significant benefits, including cost savings, improved accuracy, and the potential to reduce biases like redlining in urban development.

The integration of the metaverse into construction and real estate is setting the stage for a more connected, efficient, and inclusive approach to the sector, with far-reaching implications for both the physical and virtual worlds.

Virtual and Augmented Building

Meanwhile, back in the metaverse, a virtual real estate boom is unfolding. Across multiple platforms, the opportunity to purchase and develop virtual land—alongside building, buying, selling, or renting properties—is attracting investors and curious explorers alike. What starts as a design in the metaverse may very well be buildable in the real world, illustrating the unique intersection of construction and real estate between these two realms.

At this intersection, the metaverse offers a dual narrative. On one side, the metaverse is being used to design properties that will be constructed in real life, known as built space. On the other side, there are entirely virtual properties that exist solely in the metaverse. Both types of properties can be showcased, marketed, and sold through the metaverse, but let's begin with built properties.

In the design and construction of built spaces, the metaverse plays a pivotal role.[41] Through digital twinning, architects, engineers, builders, construction teams, environmental experts, policymakers, owners, and other stakeholders gain a powerful tool for design and decision-making. A metaverse-based digital twin of anything from a tiny house to a large-scale development allows all involved parties to assess their stakes in the project comprehensively. This tool aids in tracking costs, assessing structural integrity, testing material choices, determining environmental impacts, and even gauging the psychological effects of the building on its future occupants and visitors.

The metaverse is also revolutionizing urban operations and service delivery. Federal agencies like FEMA (Federal Emergency Management Agency) and the Department of Transportation are developing 3D city maps to enhance disaster preparedness, transportation planning, infrastructure management, public engagement, and education. These 3D maps offer first responders—such as police, firefighters, and emergency medical technicians—valuable tools for delivering emergency assistance and responding more effectively to dangerous situations. AR, VR, and these advanced 3D maps enable these entities to navigate complex scenarios and model the impacts of various factors, such as weather conditions or hazardous substances on pedestrian and vehicle traffic, helping determine how best to utilize evacuation routes and medical services in case of an emergency.

Traditionally, the visualization of such factors was limited to reports, tables, or two-dimensional charts. In the metaverse, however, these static reports are being replaced by dynamic simulations that blend data with empathy. This immersive approach allows users to experience the real-time consequences of decision-making, combining art and science in ways that help citizens and urban planners develop more effective plans, policies, and guidelines for various conditions. The result is a heightened ability for agencies to respond swiftly and confidently.

When developing these metaverse tools or designing urban applications, it is crucial to involve a broad range of designers and community stakeholders. Their input can prevent unintentional biases and ensure that critical, potentially lifesaving elements are incorporated. This includes recognizing and appropriately responding to diverse faces, behaviors, and communication styles, as well as adapting rescue protocols to accommodate individuals with varying degrees of mobility. Without input from a wide spectrum of users and experts, these tools risk overlooking key considerations that could impact accessibility and safety.[42]

The Metaverse Is Reshaping Real-World Building

The digital building boom, where AR and VR technologies are increasingly integrated into real-world construction, is still in its early stages, but it holds transformative potential for the industry. By enhancing efficiency, quality, and sustainability, these digital innovations can help create a more sustainable built environment.

AR and VR technologies are being used across the entire lifecycle of building design, construction, maintenance, and operation. These applied technologies streamline processes, reduce the need for costly changes, and ultimately save money. For instance, VR allows architects and engineers to immerse themselves in a virtual environment, identifying potential design flaws before construction even begins. This proactive approach ensures that contractors can install building systems with greater accuracy, reducing errors and improving the overall quality of the build. Additionally, virtual simulations of different materials' impacts on the indoor environment enable more informed decision-making, leading to buildings with a smaller environmental footprint.

The metaverse's capacity for collaboration is particularly powerful, enabling codevelopment on scales ranging from small neighborhoods to entire cities. Immersive urban development resources allow planners to avoid common pitfalls and costly mistakes. Virtual models provide real-time insights into how complex factors might interact, helping to foresee both desirable and undesirable outcomes. While creating a digital twin of a project may involve significant upfront costs, tweaking the digital model is far less expensive than making real-world adjustments post-construction. As the industry progresses, experts predict that virtual collaboration and design will become integral from start to finish.

Despite the potential, challenges are inevitable as this is still uncharted territory. Companies are already engaging with the metaverse across various stages of the building process, starting with existing AR applications and moving toward more immersive experiences. They design and test in the metaverse and then execute in the real world. However, there are also instances where what is designed in the metaverse remains within that virtual realm, serving as a testbed for ideas that may not yet be ready for physical implementation.

VR also supports all aspects of planning and coordination in construction, reducing human error and boosting efficiency. Problem-solving becomes more accessible when builders can invite experts into the metaverse to review challenges in real time.

Interior design is another area where the metaverse is making a significant impact. Designers can now create multiple options for a space and allow clients to fully experience these designs before making a decision. This immersive approach goes beyond traditional sketches and swatches, enabling clients to interact with potential designs in a highly realistic environment, complete with visual, auditory, and even tactile feedback through haptic VR gloves.

When it comes to marketing and selling properties, the metaverse is poised to revolutionize the process. During the peak of Web 2.0, virtual house tours became a powerful tool in the real estate market, featuring high-resolution images, videos, and user-controlled walk-throughs. These tools are still valuable today, but the metaverse offers an even more immersive way to showcase properties. With a VR headset, potential buyers can tour properties, interact with real estate professionals, and gather essential information in a way that feels almost as tangible as visiting the property in person. This not only enhances how properties are marketed but also changes what is being offered and sold, as the metaverse continues to evolve the real estate landscape.

The Virtual Property Boom

Get ready for the digital real estate rush. While real-world property markets may be cooling, virtual land in the metaverse is heating up, with plots selling for millions of dollars and with approximately 2 billion USD worth of land changing hands in 2022.[43] Investors, startups, celebrities, and major corporations are diving in, eager to claim their stake in what many see as the next big frontier. The metaverse is poised to revolutionize real estate and construction, particularly in the realm of virtual development.

There are several reasons why people are flocking to invest in virtual property.[44] For one, they anticipate that the value of these digital assets will appreciate as the metaverse becomes more mainstream. Virtual real estate offers opportunities for generating income through leasing, development, or personal use. Despite the inherent risks, the virtual property boom underscores the growing excitement and belief in the metaverse's potential.

While real-world real estate faces uncertainty, the virtual real estate market is thriving. For example, a plot in Decentraland recently sold for 2.4 million USD—a clear sign that the digital land rush is well underway.[45] Unlike traditional real estate, these properties consist of bits, bytes, and blockchain rather than bricks and mortar. Early investors are still exploring the metaverse, applying tried-and-true real estate strategies from the physical world, such as buying, building, and renting out spaces. Just as in real life, factors like location, use, and community will influence value and drive prices.

However, investing in virtual real estate comes with its own set of challenges and opportunities. The market is still young and experimental, making it both manageable and unpredictable. Investors should approach with caution, much like the early days of online real estate listings in the 1990s when access to MLS data disrupted the traditional broker model. The current landscape is competitive, with platforms like The Sandbox leading the market, accounting for 70 percent of 2021's metaverse transactions, followed by Decentraland and Voxels, 24 percent and 6 percent respectively.[46]

As more players enter the metaverse, it's still uncertain which platforms will dominate.

In the meantime, conventional real estate investment models can still be applied to generate income in the metaverse. However, investors must stay informed about market trends and avoid purely speculative ventures. It's crucial to secure the necessary expertise in understanding the nuances of both real estate and the metaverse, as well as cryptocurrency. Transactions in the metaverse are often fast-paced and governed by smart contracts, making it easy to make—or lose—money.

For commercial real estate, the metaverse offers a two-way flow of opportunities. Individuals may want their real-world homes replicated in the metaverse, protected by NFTs, or they might seek to build a virtual home in the real world. In 2022, construction was among the top sectors investing in the metaverse, accounting for 5 percent of the total investment—signaling that the industry sees the potential for collaboration, design, planning, and execution in this new digital frontier.[47]

The bond between commercial real estate and the metaverse is only set to strengthen as Web 3.0 and metaverse technologies become more stable. Just as commercial real estate professionals were early adopters of the internet, using it to elevate their marketing in the '90s, they are now poised to harness the tools of the metaverse to enhance sales and services. From immersive 3D home tours to virtual planning and digital twinning, the metaverse offers a full spectrum of applications that can transform everything from large-scale urban development to detailed interior design. As the metaverse gathers momentum, it's clear that the synergy between commercial real estate and virtual spaces will continue to grow, reshaping the industry in profound ways.

Media and Entertainment

In another industry and economic sector, the metaverse has already built momentum. Virtual concerts, shows, and entertainment events have gained significant traction in the metaverse, offering promoters and organizers a revolutionary way to reach audiences globally.[48] The Travis Scott virtual concert in *Fortnite* is a prime example of how the metaverse can create immersive experiences that draw massive crowds, with over 27 million viewers tuning in.[49] The metaverse offers unique advantages, such as allowing fans to enjoy a more personalized and interactive experience, transforming them from passive observers to active participants in performances and stories.

Media and entertainment companies are capitalizing on these capabilities by developing and marketing tiered experiences that cater to different levels of immersion and engagement. As attendees explore these virtual spaces, they encounter a blend of entertainment and marketing, where advertising becomes a seamless part of the experience. This integration of immersive content with marketing and sales creates new revenue streams and deepens consumer engagement.

Telecommunications companies are also stepping into the metaverse, offering content and recreational experiences that keep users connected while exploring new opportunities.[50] Vodafone, for instance, provides virtual concerts, stargazing nights, and interactions with celebrities, enhancing the entertainment value of its service. South Korea's SK Telecom launched its metaverse platform ifland in 49 countries in 2022, branding itself as "The New Way of Socializing" with offerings like K-pop content and interactive fan experiences. Deutsche Telekom's upcoming T-verse (owned by T mobile) and NTT Docomo's XR World are set to deliver a variety of social, educational, and entertainment experiences, including virtual museum tours and historical site visits.[51]

Immersive and interactive media content, where users have control over their experiences, is expected to hold significant value in the metaverse.[52] Companies are keenly aware of this as they carve out their spaces in this new digital frontier. Warner Bros.' Readyverse Studios and Netflix's virtual theaters are examples of how major players are preparing to offer interactive experiences that allow users to engage with their favorite characters and stories in ways that were previously impossible. The success of virtual events in the metaverse is already evident. Ariana Grande's 2021 Rift Tour on *Fortnite* concert delivered a groundbreaking spectacle, while Travis Scott's concert captivated over 27.7 million viewers.[53] The ability of the metaverse to remove geographical barriers and bring together individuals from various locations for shared experiences is creating a new wave of opportunities for event organizers, promoters, and marketers. As the metaverse continues to evolve, its impact on entertainment and media is set to be profound, offering new ways for users to connect, engage, and be entertained.

10 Telecoms, Tech, and Transformation

Telecommunications, or telecoms, play a central role in both the internet and the metaverse, laying the groundwork upon which these digital worlds are built. Telecom companies are responsible for developing the high-speed, low-latency connections essential for the metaverse's immersive experiences. They also contribute to creating the security and privacy features that will protect users in these virtual environments. As technology companies continue to innovate, dazzling consumers with ever more sophisticated augmented reality (AR) and virtual reality (VR) products, telecoms ensure that these advancements are accessible to the world by bringing the content and services to a global audience.

Both technology companies and telecoms have significant roles in promoting inclusivity within the metaverse. Technology companies can do so by making AR and VR equipment more affordable, while telecoms extend networks into remote and underserved communities, ensuring broader access to these emerging technologies.

The metaverse is also poised to have a transformative impact on energy consumption and management. It will demand substantial energy resources, but it also offers new ways to optimize and manage energy use. Energy professionals will be able to create virtual models that simulate energy consumption, allowing for the identification and implementation of more efficient practices. The metaverse can also serve as a platform for raising awareness about energy consumption and for innovating and scaling renewable energy options.

In manufacturing, the metaverse is already driving significant changes. Companies are using digital twins and virtual simulations to design, build, and test products more efficiently. In aerospace, Lockheed Martin and Boeing are refining engineering processes through digital twins, while Ford,

BMW, and Rolls-Royce are pioneering virtual assembly lines to enhance automotive production. Medical equipment manufacturers are simulating real-world conditions in virtual spaces to improve product testing. Siemens, GE Aerospace, and GE HealthCare (formerly known as GE) are utilizing immersive simulations to innovate industrial equipment, streamlining processes from inception to execution.

Across various industries, from automotive to aerospace, companies are embracing these digital tools to revolutionize manufacturing with precision and agility. For instance, Airbus and Rolls-Royce are conducting complex aircraft tests in virtual environments, reducing risk and resource expenditure. In the electronics sector, Samsung is rapidly exploring design variations through virtual prototypes, while India's Bajaj Auto is envisioning the next generation of automotive engineering in VR environments. Organizations like CSIR in South Africa and Embraer in Brazil are applying metaverse technologies to break new ground in scientific research and aeronautics, demonstrating that the future of manufacturing is as expansive as the virtual worlds being created. The metaverse is not just changing how we connect and entertain but also how we innovate, design, and produce across multiple industries.

Technology and Telecommunications

The metaverse is becoming the next frontier for communication, and people are ready to meet there. While new metaverse users may be slightly behind gamers and *Second Life* enthusiasts who have long interacted through avatars, the interest in moving beyond traditional video calls, conferencing, and social media is growing rapidly. This makes the metaverse a natural extension for telecom companies, as communication is at the core of these virtual worlds. However, the metaverse also presents a disruptive challenge: what is successful now in traditional media and telecommunications may not necessarily translate into success in the metaverse.

Telecoms and media companies do have several advantages. They have large user bases, existing content libraries, and expansive high-speed networks, all of which are crucial for a seamless metaverse experience. However, these companies must be ready to adapt and innovate, as the metaverse demands new approaches and strategies. Leaders in telecoms have a unique set of options to influence the quality of networks and, by extension, the overall user experience in the metaverse.

These leaders also have the opportunity to experiment with content and collaborate with other businesses to explore and expand the possibilities of the metaverse. By leveraging their existing assets and exploring new partnerships, telecoms can help shape the metaverse into a vibrant, interactive space that meets the needs and expectations of users who are eager to dive in. The metaverse represents both a challenge and an opportunity for telecoms, and those that are willing to innovate and adapt will be best positioned to succeed in this new digital landscape.[1]

Multiple Roles

Telecommunications companies are playing a pivotal role in the development and expansion of Web 3.0 and the metaverse, enabling continuous, immersive engagement with low network latency.[2] The telecommunications industry is crucial in building the network infrastructure that powers the speed and reach of the metaverse. This represents a key area of innovation and growth for the industry.

Many telecoms are heavily involved in the rollout of 5G networks, which will bring the metaverse to less connected and more remote areas. In the US, companies like AT&T, Verizon, and T-Mobile are investing in 5G and edge computing, focusing on enhancing mobile and low-latency networks.[3] In Europe, Orange and Vodafone are similarly advancing 5G and cloud computing, while the UK's BT Group is expanding 5G and fiber optic networks.[4] In Asia, major players like China Mobile, NTT Docomo in Japan, and SK Telecom in South Korea are not only building 5G infrastructure but also collaborating with leading metaverse companies such as Tencent, Alibaba, Sony, and Samsung.[5] Africa's MTN Group is focusing on 5G and rural connectivity, while Australia's Telstra, Optus, and TPG Telecom lead the way in their region.[6] These companies are also forming partnerships with Meta, Microsoft, and other tech giants, signaling their interest in more than just providing the metaverse's backbone.[7] *TechCrunch* and *The Verge* were the first to report that Meta is planning to build a world-spanning fiber optic subsea cable for its own use, creating a new network spanning over 40,000 kilometers (about 24,850 miles) around the globe.[8]

The installation and rollout of 5G networks has opened up new possibilities, particularly in sectors like health care, education, and manufacturing. The superior speeds promised by 5G are expected to deliver superior experiences across these industries, offering significant economic opportunities. The potential for revenue generation is tied to the digital re-creation of cities, towns, and landmarks, the development of immersive retail and entertainment venues, and the expansion of remote jobs and affordable health care delivery.

By the end of 2022, 5G networks were established in 72 countries and nearly 2,000 cities, with numbers always on the rise.[9] Much of this infrastructure has been funded by mobile carriers, which anticipate substantial returns as AR, VR, and other emerging technologies become more widespread.[10] The metaverse, deeply intertwined with 5G, will likely reach users in the same way the internet does today. With mobile phones accounting for 56.9 percent of online time and 60 percent of global web traffic, and 92 percent of internet users accessing the web via mobile devices, the telecommunications industry is positioned to play a central role in the metaverse's growth and success.[11]

Given these investments, expectations, and trends, businesses can begin making strategic decisions that will enable them to thrive in the rapidly approaching future.

No Dominant Player, Yet

The metaverse is emerging, but it hasn't yet seen the rise of a dominant player. This could be due to its global nature, which challenges any one platform's ability to claim billions of users. The situation is reminiscent of early phases in technology adoption over the past few decades. For example, during the early days of personal computing, the term "IBM-compatible" dominated the market, signifying the company's early lead. Similarly, Netscape once held a significant position in the early internet era, though it eventually faded, and its domain now redirects to AOL.

With the metaverse, we're still in the early stages, and the landscape is open for leadership. Innovators and leaders are searching for models that can guide their strategies. The vision of a global, interoperable metaverse—where users can seamlessly "world hop" across various platforms—highlights the importance of building inclusively. This involves considering language, culture, and other differentiating factors to ensure broad accessibility and appeal.

"World hopping" within the metaverse is akin to exploring different worlds and offers vast opportunities for "world shopping." For instance, XR World is a metaverse platform that offers users virtual tours of museums and historical sites and provides various educational and training experiences at multiple price points.

Moreover, several telecommunications companies are forming partnerships to build services across diverse sectors, including consumer goods, industrial goods, and manufacturing. These partnerships range from platforms focused on entertainment to those helping companies leverage digital twins in their research and manufacturing processes. As these collaborations take shape, they will likely play a critical role in shaping the future of the metaverse and its potential to become a truly global and inclusive digital universe.

Partnering with a Purpose

Verizon, Telefónica, Vodafone, SK Telecom, and Orange are among the telecom giants making significant strides in the metaverse. As a key player in the expansion of 5G and future 6G networks, Verizon has naturally aligned itself with Meta, focusing on enhancing network speed and quality, particularly in cloud rendering and reducing latency—critical elements for a seamless metaverse experience.

Telefónica has taken a unique approach by partnering with automotive parts manufacturer Gestamp to establish a 5G-connected factory in Barcelona. This facility uses virtual reality to run complex simulations, significantly reducing the time required for testing and development. Meanwhile, Orange has also been active in the metaverse. In 2022, its Luxembourg branch introduced a metaverse-based resource aimed at teaching digital skills. Later that year, Orange Spain launched a metaverse store, marking its foray into virtual retail.

These initiatives by telecom companies are not just about generating revenue; they also aim to extend new services to customers, build virtual communities, and explore the latest innovations in interaction and communication. Each company is carving out its niche, focusing on different, non-overlapping areas of the metaverse.

Telecoms are projecting that the metaverse will become as integral to daily life as the internet is today. In addition to Verizon, companies like SK Telecom and China Mobile are actively building platforms that blend virtual and real-life experiences. According to GSMA Intelligence, a staggering $720 billion was invested in establishing 5G networks between 2021 and 2025. By 2030, the successful implementation of AR and VR applications in the metaverse could generate $712 billion in revenue for telecoms, providing them with the opportunity to recoup their substantial investments.[12] This underscores the significant potential the metaverse holds for the telecom industry as it continues to evolve and expand.

Observed Trends

Cashing in on the metaverse requires telecoms to rethink their strategies.[13] Their initial forays into the metaverse have allowed them to safely test the waters and build brand awareness, but as they venture deeper in pursuit of a larger share of the metaverse economy, they will face new risks and challenges. One key challenge is the need for extensive partnerships that compel telecoms to pay closer attention to the ecosystems influencing the viability of metaverse products and services.

Success in the metaverse often depends on the seamless integration of devices, networks, and services. For instance, a faster network can precede the launch of an advanced device, but if that device produces the same results on an unimproved network, it can lead to setbacks for both the device manufacturer and the telecom provider. Consumers expect a flawless experience, where devices and networks work in harmony. Any disruption—such as a dropped virtual call, a sudden offline experience in a metaverse museum, or glitches during a virtual tour—will result in dissatisfaction, regardless of whether the fault lies with the device or the network. Everyone in the ecosystem risks negative repercussions. Telecoms have four distinct areas of opportunity to focus on to produce results in the metaverse:

Personalization and Individual Expression

Programmable and Interactive Worlds

Enhanced AI Integration

Innovation and Advanced Computing

The first trend involves those aspects of the metaverse we may think of as personal. The metaverse allows individuals and businesses to create and project their personas in a digital space, supported by artificial intelligence (AI) and other technologies. Telecoms can play a critical role in enabling these personal experiences by providing the connectivity and infrastructure necessary for users to establish their digital identities, test business models, and refine strategies within the metaverse.

This leads into the second opportunity for telecom centers related to the programmable nature of the metaverse. The metaverse offers the ability to replicate and customize a digital equivalent of the world through technologies like digital twinning, ambient computing, AR/VR, and smart materials. Telecoms are increasingly involved in the connection, experience, and material layers of the metaverse.[14] As more devices and users become connected, privacy concerns will grow. Managing these risks will be crucial for telecoms, as their ability to safeguard users will influence their value in the metaverse ecosystem. Managing these risks to users will help shape outcomes for communications service providers (CSPs) and determine value. The willingness of a tech innovator to partner with a CSP will be influenced by the CSP's capacity to safeguard individuals and businesses on its networks. Device manufacturers, content creators, and service providers (banking, finance, e-commerce) take on significant risks as they offer hardware and transactions that could expose users to breaches of privacy. CSPs have to work in partnership with these entities to facilitate and maintain security.

The third and fourth trends involve enhancing AI integration, innovation, and advanced computing. AI plays a significant role in online customer service experiences, often handling initial interactions before human involvement. Telecoms have a vital role in facilitating secure and seamless metaverse experiences by supporting AI-driven services. As AI becomes more sophisticated, telecoms will need to ensure that these technologies are secure and reliable, enhancing consumer trust and engagement.

The metaverse is still evolving, with ongoing refinement and improvements. Telecoms are well positioned to participate in the next waves of innovation, including the potential development of quantum-based processors. Their role extends beyond providing connectivity—they are strategic partners in the development and implementation of new technologies and policies within the metaverse, ensuring that it is built and tested with both consumer satisfaction and security in mind.

Energy

The metaverse is quickly becoming a transformative tool in the energy sector, enabling virtual worker training, asset management, and customer engagement in ways that were previously unimaginable. For instance, the US Department of Energy has adopted the metaverse to train workers at nuclear power plants, ensuring they are prepared for real-world scenarios through immersive VR experiences. Companies are using digital twins to monitor and manage power plants, pipelines, and transmission lines with greater efficiency, allowing for early detection of potential issues. Additionally, private businesses are leveraging VR to create virtual showrooms where customers can explore new energy products and services and access real-time data on their energy consumption and costs. Meta, for example, is collaborating with energy companies to develop metaverse experiences that educate customers about renewable energy and energy management.

However, the transition to the metaverse presents a unique opportunity for the energy sector to address its inclusivity challenges.[15] Currently, women make up only 32 percent of the global energy workforce, and racial and ethnic minorities account for just 22 percent. Leadership roles within the sector are even more skewed, with women holding only 19.4 percent of these positions in 2023, despite years of efforts to improve gender diversity.[16] As customers and consumers interact with energy companies in the metaverse, these disparities may become more visible, highlighting the need for more inclusive practices. The metaverse could serve as a platform to address and reduce these gaps, fostering a more equitable industry.

Moreover, customer engagement in the metaverse should also increase awareness of the energy consumption associated with AR, VR, and other immersive technologies. The energy demands of the metaverse, especially when considering activities like bitcoin mining, are substantial and have sparked debates about their environmental impact. The metaverse presents both a challenge and an opportunity in this regard.[17] While it requires significant energy inputs, it also offers a powerful medium for exploring new energy sources and designing strategies to reduce carbon emissions. For example, virtualized travel and tourism within the metaverse can reduce the need for physical travel, thereby cutting down on transportation-related emissions, which account for 30 percent of the world's energy demand and 23 percent of carbon dioxide emissions from fuel combustion.[18]

The key challenge is to design a metaverse that is climate-positive—one that consumes energy efficiently while contributing to the reduction or reversal of carbonization. Achieving this goal will require more than just transitioning to renewable energy sources; it will demand innovative approaches to energy use within the metaverse itself. As with real estate and tourism, the dynamics of energy use will move fluidly between the real world and the virtual world, creating both challenges and opportunities.

The metaverse has the potential to enhance our ability to discover and prototype more efficient energy solutions through digital twinning and other advanced technologies.[19] However, it also risks becoming a significant energy consumer, exacerbating environmental issues rather than alleviating them. Innovating the metaverse to focus on energy efficiency and minimizing environmental impact is essential. The energy sector, along with other industries, must recognize the dual role of the metaverse: as both a tool for finding sustainable energy solutions and a contributor to the current energy consumption problem. Balancing these roles will be crucial for the future of the metaverse and its impact on the environment.

The Metaverse and the Energy Company

Energy companies are increasingly recognizing the metaverse as a powerful platform to build and strengthen their brands, particularly in markets where customers have the option to choose their providers. The immersive nature of the metaverse offers a unique opportunity for these companies to cultivate positive relationships with customers by enhancing their brand recognition and loyalty. By leveraging the metaverse, energy companies can educate their customers on energy efficiency, emergency preparedness, and other essential topics through interactive and engaging experiences.

One of the most significant advantages of the metaverse as a teaching tool is its ability to virtualize a wide range of customer interactions. For example, routine tasks like bill paying and home energy assessments can be conducted in a fully immersive environment. Customers can interact with avatars or bots representing company employees, making the experience more personal and effective. The use of 3D virtual renderings of customers' homes to demonstrate energy usage and potential waste can be a compelling educational tool, offering a more persuasive and impactful way to encourage energy-saving behaviors.

This approach not only helps customers make informed choices about their energy consumption but also positions energy companies as proactive partners in their customers' efforts to reduce waste and improve efficiency. By providing information in a medium that is both engaging and easy to understand, energy companies can drive meaningful action that transforms energy usage patterns.

Beyond customer engagement, the metaverse presents direct opportunities for energy companies to upskill their workforces. The platform is ideally suited for routine, hands-on training in a sensory-rich virtual environment. Employees can practice complex tasks repeatedly in a safe and controlled setting, improving their skills and confidence. This is particularly valuable for training in dangerous procedures and operations, such as handling downed powerlines or preventing disasters in oil and gas exploration. In these cases, the metaverse allows for the simulation of hazardous scenarios without the risks associated with real-life training, ensuring that employees are better prepared for any eventuality.

Overall, the metaverse offers energy companies a versatile and effective platform for both customer education and workforce development. As the metaverse continues to evolve, it will likely become an increasingly integral part of how energy companies interact with their customers and manage their operations.

Finding New Energy

The metaverse offers the energy sector innovative ways to explore and model complex systems, making the business of energy safer and more cost-effective.[20] Digital twinning and industrial replicas allow companies to simulate and assess risks, optimize operations, and improve efficiency without the need for costly and time-consuming real-world prototyping. This capability introduces a level of efficiency that can significantly reduce carbon emissions, as it minimizes the need for physical trials and reduces waste in the development process.

However, powering the metaverse comes with its own set of challenges. While the digital world may take up no physical space, it leaves a sizable carbon footprint. The infrastructure required to support the vast array of servers and devices accessing the metaverse consumes massive quantities of energy. As advancements in speed, reduced latency, and the introduction of new content and services continue to evolve, the energy demands of the metaverse are expected to grow.

This duality presents a critical challenge for the energy sector: while the metaverse can contribute to emission reductions through more efficient modeling and planning, it also has the potential to exacerbate energy consumption and, consequently, carbon emissions. The balance between these opposing outcomes will depend on how effectively the energy sector can innovate and implement strategies that optimize energy use within the metaverse, while continuing to develop sustainable energy solutions in the real world.

Gobbling Up Energy

The introduction of new, faster data centers by Meta (then Facebook) between 2018 and 2019 offers a glimpse into the energy-intensive demands of the digital age. During this period, Meta's power consumption surged from 3.4 to 5.1 terawatt-hours—a staggering increase—yet it managed to cut greenhouse gas emissions by 59 percent.[21] This paradox highlights the double-edged nature of technological advancements: while new technologies can reduce emissions through improved efficiency, they also come with significant energy costs.

The metaverse, with its promise of virtual tourism, remote work, and other digital activities, presents multiple opportunities to further reduce emissions by decreasing the need for fuel-based travel. However, the energy demands of the metaverse are expected to be immense. AI processing, which is integral to the functioning of the metaverse, is particularly energy intensive. A study by the University of Massachusetts at Amherst revealed that training certain large AI models could result in more than 626,000 pounds of carbon dioxide emissions, a stark reminder of the environmental cost of advanced computing.[22]

Compounding this concern, Intel experts project that the global computing infrastructure will need to become 1,000 times more powerful just to support the metaverse at scale.[23] With the potential to support nearly 5 billion users, the energy requirements could be astonishing, raising critical questions about sustainability and the environmental impact of this digital frontier.

In addition to the environmental challenges, power fluctuations and failures in such an energy-intensive environment could create vulnerabilities for bad actors. These issues may invite data breaches or disrupt virtual locations and communities, potentially leading to chaos in a world that is increasingly reliant on digital interactions. As the metaverse evolves, balancing the benefits of virtual experiences with the environmental and security challenges it poses will be a critical task for leaders in technology, energy, and governance.

At the Energy Crossroads

The metaverse, with all its potential, presents a complex dilemma for the energy sector.[24] On the one hand, it offers unprecedented opportunities for innovation, consumer engagement, and efficiency in business operations. On the other, it poses significant challenges in terms of energy consumption and environmental impact. As the virtual universe draws power from the real one, the strain on global energy resources could be immense, leading many to argue that the metaverse may do more harm than good.

The energy demands of the metaverse are undeniable. As businesses and consumers increasingly engage in activities within this digital realm, the pressure on power grids and the overall environmental footprint will likely grow. Large tech companies like Meta, Microsoft, and Apple have already recognized this challenge, committing to renewable energy sources and aiming for net zero emissions in their value chains by 2030.[25] However, these efforts alone may not be enough to offset the enormous energy consumption required to support a fully realized metaverse.

One of the critical issues is the efficiency of the technology powering the metaverse. High-resolution video, immersive audio, AR, and VR experiences require substantial energy. While companies like Meta are making strides with products like the Oculus Quest, which offers a more sustainable AR option, the broader challenge remains. The energy required to sustain a global metaverse far exceeds what was necessary for the internet or global telephony, making it essential for innovators to focus on creating more efficient hardware and software.

Moreover, consumer behavior plays a crucial role in this equation. The current internet already experiences peak traffic times, typically between 6:00 p.m. and 11:00 p.m., when social media, gaming, and streaming activities surge. These peak hours could strain power grids even further as the metaverse becomes mainstream. Encouraging more disciplined energy use, developing devices that consume less power, and expanding renewable energy sources are necessary steps to mitigate these stresses.

The environmental impact of emerging technologies, particularly in the realm of cryptocurrency, further complicates the picture. The carbon emissions associated with bitcoin mining, for example, have skyrocketed (from 0.9 tons in 2016 to 113 tons in 2021), drawing criticism for their environmental toll.[26] The yearly carbon footprint is comparable to that of the county of Greece, and its 75.4 terawatt-hours of electricity consumed exceeds that of Portugal and Austria, at 48.4 and 69.9 terawatts, respectively.[27] If the metaverse follows a similar trajectory, it could exacerbate existing challenges rather than alleviate them.

For the metaverse to succeed without causing significant harm, a multifaceted approach is required. This includes technological innovation focused on energy efficiency, consumer education on responsible energy use, and a broader commitment to renewable energy. Without these efforts, the metaverse may struggle to find a sustainable path forward, risking instability and environmental degradation. As we move toward a more immersive digital future, the choices made by tech companies, energy providers, and consumers alike will determine the metaverse's impact on our planet.

Making Changes

While the decisions made by tech companies and energy providers will have the greatest impact, individual choices still play a role. Shifts in consumer behavior—such as opting for energy-efficient devices, supporting sustainable digital practices, and advocating for responsible innovation—can contribute to broader systemic change. Achieving this requires a delicate balance: guiding and educating consumers in ways that encourage sustainable practices without making these changes feel forced or overly restrictive. Behavioral science offers valuable insights into how to design these nudges—small, voluntary actions that can lead to significant positive outcomes—without alienating consumers.

Behavioral Science and Consumer Engagement

To encourage consumers to adopt energy-saving practices, it is crucial to craft messages and design choices that resonate with their values and preferences. For instance, rather than imposing restrictions on binge-watching or high-definition streaming, content providers could frame energy-saving options as enhancements to the user experience. Offering incentives for choosing more energy-efficient options, like watching content in standard definition or engaging in shorter, more interactive experiences, can help consumers feel empowered in their choices rather than constrained.

The concept of nudging is particularly effective here. A nudge gently guides consumers toward a desired behavior without removing their freedom to choose otherwise. For example, platforms could default to energy-saving settings with the option to opt out, or they could offer rewards for consistent use of eco-friendly options. This approach highlights the personal and collective benefits of these behaviors, fostering a sense of participation in a larger, positive movement.

Promoting Sustainable Tech Use

Another critical area where behavioral science can play a role is in promoting the reuse and recycling of digital devices. The tech industry has long thrived on the appeal of the new and the cutting-edge, but this mindset contributes to a growing problem of e-waste. By shifting the narrative to emphasize the value and impact of reusing and recycling devices, businesses can encourage more sustainable consumer habits.

Incentives, such as discounts on future purchases or contributions to environmental causes, can make the choice to recycle or buy preowned devices more appealing. Additionally, integrating information and resources for responsible recycling directly into the purchasing process can make it easier for consumers to follow through. A well-designed nudge might involve a simple, clear message at the point of sale, reminding consumers of the environmental benefits of recycling and offering convenient ways to do so.

Shared Accountability and Partnerships

Achieving carbon reduction in the metaverse requires a collective effort from all stakeholders—businesses, consumers, and energy providers alike. Large corporations can lead by example, not only by transitioning to renewable energy sources themselves but also by promoting sustainable behaviors among their customers. For instance, metaverse businesses could incorporate environmental education into their platforms, helping users understand the impact of their digital activities and offering suggestions for more sustainable practices.

Content providers, device manufacturers, and energy companies can collaborate to create a more eco-friendly metaverse. For example, recommendations for lower-energy consumption could be built into streaming services, or energy-saving modes could be made the default setting for new devices. These efforts could be reinforced with public awareness campaigns that emphasize the importance of individual actions in contributing to a larger environmental goal.

A Net Positive Vision for the Metaverse

The metaverse has the potential to be a net positive for the environment, but realizing this potential will require careful planning and a commitment to sustainable practices. By leveraging behavioral science to encourage eco-friendly behaviors, promoting the reuse and recycling of digital devices, and fostering partnerships between businesses and consumers, it is possible to create a metaverse that contributes to carbon reduction rather than exacerbating the problem.

As we move forward, it will be essential to continue exploring innovative ways to reduce the energy demands of the metaverse while also educating users about the environmental impact of their choices. By making sustainability an integral part of the metaverse experience, we can ensure that this new digital frontier supports a healthier planet for all.

Manufacturing

The manufacturing industry faces a looming crisis as 2.7 million baby boomers are set to retire by 2025, leaving an anticipated 2 million vacant jobs. This gap will likely widen as more workers age out of the workforce. Compounding the problem, industries like construction and mining, which have historically lacked diversity in gender and demographic participation, continue to struggle to attract a broader pool of talent. Outdated perceptions and misconceptions about these fields further deter potential workers. Women in particular leave manufacturing roles at higher rates than men. A 2015 study revealed that while two-thirds of women in manufacturing would choose to stay in the sector if restarting their careers, many still leave due to unsatisfactory workplace relationships, limited promotion opportunities, and pay disparities.[28] With 70 percent of women perceiving a gender pay gap and earning only 83 percent of what their male counterparts do, it is unlikely that women alone will resolve this crisis.[29] Instead, the metaverse offers a promising solution that may inspire a new generation to entire the field.

The metaverse is well suited for manufacturing, with AR providing real-time visual overlays to help operators control robots accurately and efficiently along precise paths or carry-out step-by-step instructions for complex tasks. VR creates immersive training simulations that allow operators to learn how to handle manufacturing robots without risk, from any location. This use of game-like technology and the flexibility of remote work holds promise for attracting younger generations to fill the roles vacated by retiring baby boomers.

AR and VR tools adapt to a wide range of aspects of manufacturing, including product design and development, production planning, quality control, maintenance, safety, collaboration, customer engagement, training, and R&D. Even as human labor is lost, in the case of baby boomers retiring, the intersection of robotics and the metaverse may provide a viable solution. Manufacturers can establish virtual factories where workers collaborate with each other and customers globally. Already, the metaverse has improved safety, increased productivity, and reduced costs in the manufacturing industry.[30]

Leading companies are leveraging the metaverse in their operations. Boeing uses AR to train employees on aircraft assembly, improving quality and reducing assembly time. Ford employs VR to train assembly line workers, allowing them to practice in safe, controlled environments. Volkswagen reaches a broader audience and boosts sales through virtual showrooms where customers can browse and purchase vehicles. Partnerships with companies like Nvidia, Unity, and Microsoft are enhancing digital twins and virtual showrooms, with BMW, GE Healthcare and GE Aerospace, and Honeywell also using the metaverse in manufacturing, testing, training, and management capacities.

The metaverse promises to revolutionize manufacturing, offering high levels of automation, increased safety, and enhanced collaboration across distances. Virtual product design, training, safety measures, and impacts across multiple industries are already being realized. The industrial metaverse has begun its journey, utilizing digital twins, AI, blockchain, immersive simulations, advanced cloud computing, and 5G connectivity. These technologies enable large and small companies to test new possibilities, shore up supply chains, and keep decentralized collaboration on track. Employees and experts across various locations have the tools and materials needed to design and execute projects efficiently. The metaverse, therefore, represents a critical opportunity to address the looming manufacturing crisis while transforming the industry for the future.

Collaborating and Designing Anew

The metaverse is revolutionizing virtual product design by enabling the creation of highly realistic virtual prototypes.[31] These digital models can be rigorously tested in various scenarios, revealing defects and allowing for optimization before they are manufactured in the real world. This approach not only saves money but also increases product quality and ensures greater safety. The metaverse facilitates remote collaboration, enabling designers, engineers, and other industry professionals to work together seamlessly, whether in real time or asynchronously. The immersive work environment enhances the sharing of ideas and the delivery of feedback, making the design process more efficient and collaborative.

In addition to product design, it offers unique tools for virtual meetings and collaboration across industries, providing benefits that go beyond what is possible in the real world. Employees from multiple locations can gather in the metaverse, engaging in three-dimensional dialogue and problem-solving. This immersive experience fosters novel combinations of ideas and resources, representing a new frontier for manufacturing and industrial goods. The metaverse supports low-latency remote meeting simulations, where participants appear to be in the same room. For example, Meta's

Horizon Workrooms app allows users to attend and participate in meetings through their avatars, creating dynamic opportunities for collaboration.

When detailed simulations are combined with remote collaboration, the potential for innovation becomes almost limitless. Virtual models and simulations in the metaverse also pave the way for developing immersive training and testing programs, further enhancing the capabilities of industries to innovate and improve. The metaverse is not just transforming how products are designed and tested but also how professionals collaborate and solve problems, opening a world of possibilities for the future of manufacturing and beyond.

Digital Twins

A natural extension of metaverse applications in manufacturing is digital twinning, where VR simulations provide cost-effective means for collaboration and testing of new products and processes.[32] When these products are ready for the market, the metaverse offers immersive tools that can significantly enhance industrial sales. The ability for potential buyers to visit, inspect, and observe products or services in action—without being physically present—holds immense selling power. The metaverse uniquely provides this in-person, hands-on experience at a distance, allowing manufacturers to sell to customers across the globe. Through a blend of AR and VR, buyers can interact with, inspect, and even operate products, building confidence, trust, and certainty before making a purchase.

Tomorrow's Factory and Its Workers

The metaverse promises to enable and revolutionize the factory of the future, with AR and VR playing pivotal roles in all phases, from education and training to innovation and invention to sales and servicing.[33] These immersive technologies have the potential to enhance every aspect of manufacturing, warehousing, logistics, and even transportation, offering unprecedented flexibility and efficiency. As the metaverse reshapes education, it may lead to significant changes in traditional methods and models, pushing some to the brink of obsolescence. However, for businesses and industries (especially in manufacturing) this transformation presents substantial benefits. By leveraging AR and VR, companies can significantly reduce the risks of injury and optimize the learning process. Workers can be trained, verified, and certified from any location using digital twins that serve as ideal tools for educating and equipping employees with essential skills. The capacity to train and upskill the workforce using sensory-rich, hands-on instruction becomes a game-changer. Employees benefit from the ability to adapt training materials to their own learning styles and develop skills at their own pace, creating a safer, more personalized, and more effective work environment.

Onboarding and Training

In the manufacturing sector, the metaverse extends its influence beyond traditional employee onboarding and upskilling to play a crucial role in product assembly and quality control. By utilizing digital twins and remotely operated manufacturing tools, production teams can precisely build, test, and certify products with greater accuracy and efficiency. As pioneering businesses explore these possibilities, it's essential for organizations to start integrating emerging virtual tools into their corporate strategies. Moreover, VR as an instructional tool has proven to enhance overall employee skills. According to a PwC study, employees trained in virtual environments complete training up to 4 times faster than in traditional classrooms and 1.5 times faster than through e-learning.[34] They also demonstrate greater confidence, with VR learners being 275 percent more likely to act on what they've learned—significantly outperforming classroom and e-learning methods. Beyond speed and confidence, VR fosters a deeper emotional connection to training material, with learners feeling 3.75 times more engaged than in-person trainees and 2.3 times more engaged than e-learners. Additionally, VR enhances focus, with participants staying 4 times more attentive than e-learning peers. While initial costs may be higher, the PwC study found that VR proves to be more cost-effective at scale, making it an increasingly attractive solution for workforce development.

Businesses have several options for incorporating VR into their training processes: outsourcing, building, or buying. Outsourcing represents the easiest entry point, allowing companies to hire experts to design and deliver course content while focusing on their core operations. This approach ensures high-quality, professional content that optimizes learning without overburdening internal resources. Alternatively, businesses with internal expertise can build their own VR training programs. Although some may find the technology intimidating, creating VR content is increasingly accessible. Many smartphones can capture 360-degree images and footage, and apps like Panorama 360 and Google Street View offer free or affordable content creation tools. The process often involves simple drag-and-drop techniques similar to the evolution of website design from complex coding to user-friendly templates.

For standard training needs, businesses may not need to create or commission custom content. Off-the-shelf AR/VR training modules are available for topics like workplace harassment prevention, safety standards, and compliance training. These products typically allow for customization to suit specific corporate cultures and training requirements, providing a quick and efficient way to implement VR training. Leveraging the metaverse in this way enables companies to utilize cutting-edge media for workforce development, building new technological competencies that keep their teams competitive and current.

The metaverse repeatedly offers solutions across various sectors, from training and manufacturing to warehouse management and logistics.[35] Whether addressing business travel, work, meetings, or the movement and delivery of goods and resources, the metaverse applies its capabilities. In hazardous work environments, the interface of humans, robotics, and AI through AR and VR holds significant promise, particularly in fields like firefighting and law enforcement, where the metaverse can enhance the sensory connection between humans and their robotic counterparts.

Augmenting Inventory and Warehouse Management

Warehousing and logistics are set to undergo significant transformation with the integration of digital twinning in the metaverse. Businesses will gain access to more precise inventory management and logistics systems, enhancing efficiency and accuracy in moving products and people. As the metaverse continues to develop, transportation—ranging from personal vehicles to mass transit, cargo ships, and trains—may adopt a futuristic appearance. Real-time, immersive vehicle operation is already being tested, potentially leading to a highly automated and streamlined transportation industry monitored and controlled through the metaverse. Leaders in warehousing, transportation, and shipping can anticipate new opportunities as transportation hubs evolve with the increased automation of vehicles and services.

While the metaverse can't replace the physical movement and delivery of goods, it can significantly enhance the process. For example, the use of drones for package delivery could make the logistics of physical goods faster, more convenient, and environmentally friendly. Companies like Walmart and Amazon are already leveraging the metaverse to create virtual warehouses and digital twins of their supply chains. Walmart's virtual warehouse allows employees to pick and pack orders more efficiently, while the company explores new delivery technologies like drones. Amazon's digital twin of its supply chain helps identify potential bottlenecks and optimize operations. Similarly, FedEx is using the metaverse to create a virtual warehouse where customers can track their orders and view the real-time location of their packages.

These applications of the metaverse in warehousing and logistics promise to increase accuracy, reduce costs, and improve customer service, all while paving the way for a more efficient and automated industry. As the technology matures, the impact on how goods are stored, moved, and delivered will be profound, offering businesses new avenues to enhance their operations and meet the evolving demands of the market.

Warehouse Management

At the intersection of AR/VR and industrial applications, the metaverse is poised to revolutionize warehouse efficiency and management.[36] Digital twinning can be leveraged to track and manage inventories with unprecedented accuracy, while also optimizing workflows. The immersive capabilities of VR environments make them ideal for simulating processes and visualizing complex environments. This allows staff to test and refine what works best before implementing changes in real-world settings. The same principles apply to larger, more complex environments, as demonstrated by Siemens, which used a metaverse simulation to model the potential decarbonization of a neighborhood in Berlin.[37] This proof-of-concept approach highlights the practical applications of the metaverse in industrial settings.

While logistics initially seemed to pose a limit to the metaverse's capabilities, this barrier is rapidly dissolving. Although the metaverse cannot physically move boxes, the integration of robotics, digital twins, and AR/

VR interfaces can empower employees to manage, rearrange, and restock warehouses more efficiently. By creating a comprehensive digital overview of inventory locations, including items in transit, companies can manage supply chains with greater precision. The ability to simulate detailed processes holds promising potential for improving production centers and warehouse management.

The Metaverse Behind the Wheel

In the world of transit and transportation, there are aspects that seem to challenge the central role of the metaverse. While the metaverse excels in enhancing the planning, design, and management of vehicles and transportation systems, the question remains: can it actually operate a personal vehicle while the owner is occupied with other tasks? Can it drive a truck across the country or navigate a cargo ship across the ocean? If you ask this of the metaverse, the answer is likely that someone is already working on it.

Several companies are exploring the possibilities of teleoperated driving, where vehicles are driven remotely. In some cases, the metaverse plays a significant role as an enabling technology. The Holograktor, for example, markets itself as the "metaverse on wheels" and is touted as the first car designed around augmented reality.[38] This vehicle can be driven conventionally or operated remotely using a virtual reality device. The "presence" of a remote operator offers riders psychological comfort compared to fully autonomous driving systems. Brands like Nissan and BMW are integrating AR and VR into their warning and infotainment systems, though only the Holograktor is self-driven.

While the concept of metaverse-assisted remote vehicle operation is still in its infancy, the potential for further development is immense. The metaverse could play a role in creating self-driven trucks and trains that are less prone to human error, offering a safer and more efficient mode of transportation. As these technologies evolve, the metaverse is likely to continue pushing the boundaries of what is possible in transit and transportation, further blurring the lines between virtual and real-world operations.

In manufacturing, warehousing, and logistics, the metaverse offers tools to supplement existing processes. As a planning and piloting tool, it can reduce the need for extensive infrastructure. VR manufacturing options can take a product from concept to assembly, warehousing, and shipping, all supported by remote workers. Similar to the self-driving car, robotic or remote manufacturing can be facilitated by employees from any location globally. A manufacturing site operating 24/7 can be staffed by individuals across different time zones, allowing them to work during their local daylight hours without ever setting foot in the facility. This approach can boost productivity, reduce errors, and expand employment opportunities to a global workforce. Through the metaverse, people can be hired, trained, and work from anywhere, redefining the nature of work and employment in the manufacturing sector.

11 The Future of Work

From enabling remote work around the world to facilitating previously impossible collaborations to virtual resources and new work opportunities, the metaverse plans to permanently change work. As work in the metaverse becomes increasingly focused on enhanced communication, collaboration, education, and training, it promises to reshape the professional landscape permanently. Employees will find themselves reconnecting as social beings, albeit from any location they choose, while employers benefit from improved management capabilities, streamlined communication, and heightened safety measures.

The Emerging Landscape of Metaverse Work

The metaverse is swiftly becoming a new frontier for work, driven by cutting-edge technologies that redefine interaction, collaboration, and value creation.[1] As the traditional office evolves beyond physical boundaries and current remote work tools' 2D limitations, the metaverse introduces an immersive digital universe that reimagines workspaces, workflows, and employment itself.

With the increasing adoption of metaverse-based technologies, work environments are becoming more interactive, engaging, and adaptable. Central to this transformation are virtual reality (VR), augmented reality (AR), and artificial intelligence (AI), which offer novel ways for employees and employers to connect, create, and collaborate. The metaverse seamlessly integrates digital tools with human-centric work processes, enabling a more dynamic and personalized work experience.

But this new landscape presents significant challenges.[2] As businesses transition to the metaverse, they must navigate the complexities of digital transformation, including the imperative for robust cybersecurity measures, ethical considerations surrounding data privacy, and the development of new regulatory frameworks. Successfully integrating these technologies into work practices demands a multidisciplinary approach, blending insights from computer science, psychology, organizational behavior, and ethics. As we stand on the cusp of this new era, it is clear that the metaverse represents not just a technological shift but a fundamental reimagining of how we work.

To Remote and Beyond

The COVID-19 pandemic accelerated the shift to remote work, fundamentally altering employees' expectations and desires. Pre-pandemic, about 20 percent of people worked from home; during the pandemic, that number surged to 70 percent. By the end of 2020, 54 percent of people expressed a preference for continuing to work from home.[3] Surveys in 2022 and 2023 revealed a strong desire for remote work, with 65 percent of workers wanting to work remotely full time and 32 percent preferring a hybrid work environment (and 87 percent of workers stating that they take advantage of available remote options on average three days a week).[4]

While pandemic-driven remote work presented challenges in management and accountability, it also provided greater flexibility, improved work–life balance, reduced costs, and, in many cases, enhanced productivity. The metaverse can bridge the gap between employees' desire for flexibility and businesses' need for effective supervision and accountability. With a balance of trust, clear communication, and reliable productivity, remote work in the metaverse can enable teams and organizations to deliver consistent results.

The metaverse promises to bring to remote work the same opportunities and disruptions it is expected to introduce into web-based learning. It pushes the boundaries of remote work far beyond the limitations of current 2D environments. As businesses embrace the flexibility offered by remote work, they also recognize the advantages of working within a VR environment, where collaboration, productivity, and accountability can thrive.[5]

For instance, consider a global automotive company with engineering teams in Germany, production units in the United States, and a design team

The Future of Work

in Japan. With VR headsets, these teams can collaborate in real time within a virtual model of a car, discussing and modifying design elements, troubleshooting engineering challenges, and walking through assembly processes as if they were together on the factory floor, despite being continents apart. This approach accelerates the development cycle, allows immediate feedback, and ensures accountability as changes and suggestions are implemented and reviewed in a shared virtual space.

In the metaverse, all aspects of a business can be conducted virtually. Potential employees could even explore working under a temporary contract for one day at a company virtually before applying or deciding to accept a job offer. They could experience what it is like. Expensive or hazardous R&D can be shifted to the metaverse, where the possibilities are limited only by imagination. Engaging with metaverse technologies now offers businesses a competitive advantage over later adopters. Companies that failed to adapt to digital changes—like Blockbuster and Toys "R" Us—serve as cautionary tales for businesses today. Those that resist or delay embracing the metaverse may find themselves in a similar position.

However, organizations naturally have concerns.[6] While remote work has been a saving grace for many businesses, it has also faced criticism, particularly concerning issues of productivity and accountability. Some executives have ordered employees back to the office, using threats of dismissal or increased monitoring of underperforming employees. Could VR work in the metaverse offer a solution that balances the flexibility of remote work with the accountability of on-site work?

As metaverse hardware and technology continue to improve, more work is likely to migrate to this platform, replacing videoconferencing with more interactive VR meetings. The future of work in the metaverse could indeed offer the best of both worlds, driving businesses toward greater innovation and success.

Better Management and Accountability

The metaverse represents a transformative shift in remote work, fundamentally enhancing visibility and collaboration in ways that traditional setups could not.[7] In the metaverse, real-time visibility allows managers to see what employees are working on and move seamlessly from one virtual workspace to another at network speed. This capability enables early identification of potential issues and timely support, fostering a more responsive and supportive work environment.

The immersive nature of the metaverse also addresses the isolation that often plagues remote workers by enabling direct, real-time collaboration among employees. This visibility and interaction can significantly boost focus and motivation, as workers feel more connected to their team and tasks. Additionally, the metaverse offers immediate access to training and upskilling opportunities, ensuring that new hires are effectively onboarded and that existing employees remain competent and motivated in their roles.

Leading companies like Adidas, Bank of America, Meta, Microsoft, and Nvidia are already leveraging VR to enhance training, host meetings, and collaborate on projects. For instance, Meta's VR environments allow employees to learn through hands-on experiences rather than passive observation, leading to better retention and a more practical understanding of complex scenarios. Microsoft's Mesh in Teams transforms virtual meetings by replacing flat video calls with 3D spaces where participants can engage more naturally, reading body language and collaborating as if they were physically present. This creates a more engaging and personable communication experience, reducing the disconnection often felt in traditional video meetings. Nvidia's Omniverse further enhances collaboration by enabling real-time teamwork on complex projects, allowing designers, engineers, and creators to work together in a shared virtual space as if they were in the same room. This is a significant improvement over isolated and asynchronous workflows, streamlining project execution and innovation.

In addition to improving collaboration, the metaverse enhances communication between managers and employees, reducing the potential for misunderstandings.[8] The closer interaction within a virtual environment diminishes the reliance on written messaging, allowing for more immediate and clear exchanges. The presence of direct supervision and a virtual community of coworkers fosters a sense of accountability and productivity, as employees can see and be seen, creating a more connected and responsible workplace.[9]

Overall, the metaverse promises to revolutionize remote work by creating a more interactive, engaging, and productive environment that bridges the gaps of traditional remote work and enhances the overall work experience.

Revolutionizing Communication and Collaboration

The metaverse promises to revolutionize communication and collaboration, overcoming the limitations of traditional remote work. In the metaverse, colleagues can connect as 3D avatars in virtual environments, closely mimicking real-world interactions. For example, Spatial platform or Horizon by Meta allow employees from around the world to meet and collaborate in a shared virtual space.[10] This approach honors the neurodiversity of the workforce and benefits from a wider range of active perspectives.

In the health care, architecture, and construction industries, the ability to interact with and troubleshoot using 3D models offers a more efficient working method. The metaverse requires minimal real workplace hardware and can be effortlessly expanded, making it a valuable tool for modern businesses.

Improvements in Support Services and Training

The metaverse is rapidly becoming the future of customer service and support, employee training, and sales and marketing, offering transformative possibilities for businesses and employees alike. One area where this is particularly evident is in customer service. Instead of the traditional customer support call, customers and support staff can now meet in the metaverse, creating a more interactive and hands-on experience. This approach not only improves communication but also accelerates problem resolution, reducing the likelihood of repeat calls and saving businesses both time and money.

Training in the metaverse offers a range of compelling benefits that set it apart from traditional and even online learning methods.[11] The immersive nature of the metaverse, particularly through VR learning, provides a more engaging and human-centered approach to training. According to PricewaterhouseCoopers, employees who participate in VR training can be trained up to four times faster than those in conventional programs. This speed is accompanied by increased focus, emotional connection to the content, and a higher likelihood of applying the skills learned.[12]

One of the key advantages of training in the metaverse is the ability to use virtual simulations. These simulations can be customized to closely mirror real-world scenarios, preparing employees to operate machinery, respond to emergencies, or handle complex tasks. The realism of these simulations ensures that employees are well prepared for actual challenges in the workplace.

Collaborative learning environments in the metaverse allow employees to learn from peers and experts in real time, fostering a sense of community and shared knowledge.[13] Social learning tools within the metaverse encourage employees to share insights, ask questions, and collaborate on projects, enhancing problem-solving and brainstorming capabilities.

The metaverse also supports on-demand training, enabling employees to learn at their own pace and revisit material as needed. This flexibility is crucial for continuous skill development and ensures that employees can acquire new knowledge or refresh existing skills whenever necessary. Additionally, the use of gamification—incorporating points, badges, and other rewards—further enhances engagement and motivation during the learning process.

Several case studies have highlighted the effectiveness of VR-based learning, showing that it creates highly engaging and memorable training experiences. Employees trained in the metaverse retained more information and demonstrated improved professional development. This approach to training also positively impacts employee retention, as it addresses issues like skill stagnation and unclear pathways for career advancement, which are common triggers for attrition. Companies like Delta Air Lines, DHL, UPS, ExxonMobil, and Porsche are leveraging AR and VR to train their workforces, demonstrating the transformative potential of the metaverse in education and training. Through immersive, hands-on, personalized, and scalable learning experiences, employees are gaining the skills to fly planes, load and unload trucks, deliver packages safely, sell and repair products, and maintain vehicles. The metaverse is revolutionizing how businesses approach training, making it a critical tool for upskilling the workforce and driving long-term success.

Building Skills and Lifelong Learning

The advent of the metaverse heralds a new era for education and lifelong learning, where immersive and interactive experiences become central to skill development.[14] As work transitions into virtual environments, the demand for new skills will grow, and the methods by which we acquire these skills will evolve. The metaverse offers a unique platform for education, blending traditional learning with immersive, hands-on experiences that can be tailored to individual needs and learning styles.

In the metaverse, education extends beyond conventional classrooms. Virtual reality can simulate real-world scenarios, allowing learners to practice skills in a safe and controlled environment. Whether it's medical students performing virtual surgeries, engineers testing new designs, or employees undergoing compliance training, the metaverse provides a versatile and effective learning platform. This not only enhances the learning experience but also accelerates the acquisition of practical skills that are immediately applicable in the workplace.

Lifelong learning takes on new importance in the metaverse, as the pace of technological change demands continuous skill development. The metaverse supports this by offering on-demand learning resources, virtual mentorship programs, and collaborative learning environments where individuals can learn from peers and experts globally. By integrating learning into the fabric of the work environment, the metaverse ensures that education is a continuous, adaptive process that evolves alongside the needs of the workforce.

The Showcase of Bank of America

Let's look at one example of a company that has woven its employees' training into the metaverse. Bank of America's innovative use of VR in training its vast workforce of over 200,000 employees illustrates the metaverse's potential.[15] In a world driven by technological advancement, the bank's embrace of VR training signifies a forward-thinking approach to preparing its employees for the future. These immersive experiences, which include lifelike scenarios and interactive simulations, play a crucial role in developing essential skills, enhancing adaptability, and cultivating digital proficiency. The VR training is not just a novelty; it provides real-time, experience-based learning that mirrors actual situations. Trainees are exposed to sensory-rich environments, complete with sights and sounds that mimic a real bank branch. This deep engagement with training material allows employees to practice responses to real-world challenges in a controlled and measurable way, honing their problem-solving skills without the risks associated with real-world consequences.

By investing in VR training, Bank of America demonstrates a commitment to equipping its employees with the skills and confidence needed to excel in their roles, ultimately leading to improved customer service and business outcomes. These simulations act as practice repetitions, enabling new employees to quickly gain experience and confidence. The technology is used to educate staff on various scenarios, such as handling irate customers or staying calm during a robbery, proving its practical value. Additionally, VR offers virtual tours and detailed insights into the bank's branches, historical journeys, and employee benefits, showcasing a comprehensive approach to workforce development.

Bank of America's use of virtual offices, training, and customer service positions it as a leader among banks leveraging the metaverse. The company is increasingly exploring fully remote work opportunities through virtual onboarding, team building, and social events, signaling the metaverse's growing acceptance as mainstream technology. By integrating these innovative techniques into its operations, Bank of America not only nurtures a skilled, adaptable, and knowledgeable workforce but also sets a precedent for how businesses can thrive in the modern digital landscape.

A New Standard in Professional Life?

Welcome to work in the metaverse.[16] Just as tablets and laptops have become essential tools for nearly every employed person, VR headsets are poised to become a standard part of the workplace. Companies like Ford, GE, Honeywell, and Siemens are already issuing AR and VR equipment to their employees. These tools are set to revolutionize how we teach, attend meetings, collaborate, interact with HR, and even socialize with colleagues. What is currently novel will soon be commonplace, and the office will be just a headset away.

However, the metaverse workplace must also facilitate work–life balance, addressing concerns that arose during the COVID-19 pandemic when the line between work and personal life became increasingly blurred. As businesses and workers enter this new setting, they do so with both excitement and caution.[17]

Employee Well-Being

The shift to work in the metaverse brings with it profound implications for employee well-being. While the digital workplace offers new forms of autonomy and opportunities for skill development, it also presents unique challenges in maintaining social connections and a sense of belonging. The immersive nature of the metaverse can create a work environment that is either engaging or isolating (or both), depending on how it is managed.

In the metaverse, employees have the autonomy to design their workspaces, choose their virtual environments, and customize their avatars, offering a level of personalization and control that was previously unimaginable. This autonomy can enhance motivation and job satisfaction, as employees feel more in control of their work experience. Additionally, the metaverse provides a platform for continuous learning and skill development. VR and AR simulations allow employees to practice and refine their skills in a risk-free environment, leading to greater competence and confidence in their roles.

However, the transition to a fully virtual work environment also raises concerns about social isolation and the potential erosion of workplace camaraderie. The absence of physical presence can make it challenging to build and maintain strong interpersonal relationships, which are vital to employee well-being. Organizations must therefore prioritize the development of virtual spaces that foster collaboration, communication, and social interaction. By creating opportunities for employees to connect and engage with one another, businesses can mitigate the risks of isolation and promote a more inclusive and supportive work environment.

Digital Well-Being and Mental Health

As the metaverse reshapes how we work, it also redefines our approach to digital well-being and mental health. The immersive and all-encompassing nature of the metaverse can both enhance and challenge our mental and emotional states. The line between work and personal life may blur further, potentially leading to stress and burnout if not managed effectively. On the other hand, the metaverse holds promise as a tool for promoting mental health, providing spaces for relaxation, mindfulness, and social connection.

Digital well-being in the metaverse hinges on creating a balance between engagement and disengagement. The very technologies that immerse employees in virtual workspaces can also be leveraged to support mental health. Virtual environments can be designed to include spaces dedicated to mental wellness, offering VR meditations, stress-relief exercises, and social lounges where employees can connect informally. These virtual tools can foster a supportive work culture, helping employees manage stress and maintain a healthy work–life balance even in a digital world.

However, there are risks associated with prolonged exposure to virtual environments. The intensity of immersive work can lead to digital fatigue, exacerbated by the always-on nature of the metaverse. Employers must be proactive in addressing these challenges by setting boundaries, encouraging breaks, and ensuring that digital tools are used to complement rather than overwhelm employees. The metaverse offers an opportunity to redefine workplace well-being by integrating mental health support directly into the work environment, but it requires a thoughtful and intentional approach to avoid the pitfalls of overuse and burnout.

Navigating Virtual Teams and Organizational Culture

As the metaverse becomes an integral part of the work landscape, leadership must evolve to meet the unique challenges of managing virtual teams and cultivating organizational culture in a fully immersive digital environment.[18] The traditional models of leadership, grounded in physical presence and direct interaction, are being redefined in the metaverse, where leaders must navigate the complexities of remote work, digital communication, and virtual collaboration.

In the metaverse, leaders must develop new competencies to effectively manage distributed teams that operate across time zones and digital platforms. This includes mastering the use of virtual tools to communicate, motivate, and engage employees in a manner that fosters trust and accountability. The immersive nature of the metaverse offers opportunities for leaders to connect with their teams in more meaningful ways, using avatars and virtual environments to create a sense of presence and belonging.[19]

Organizational culture, traditionally shaped through shared physical spaces and in-person interactions, must be reimagined to thrive in the metaverse. Leaders play a pivotal role in setting the tone for virtual collaboration, engagement, and innovation. This shift requires a deliberate focus on creating digital environments that foster belonging, psychological safety, and meaningful participation for all employees. Rather than simply replicating traditional office dynamics, the metaverse offers new ways to cultivate an adaptable, resilient workplace culture. Leaders can enhance connectivity, encourage open dialogue, and develop interactive spaces that reflect the diverse needs of their workforce. By leveraging the unique capabilities of immersive technology, businesses can create a dynamic culture that supports collaboration, empowers employees, and ensures long-term engagement in the digital age.

Ethics and Governance in the Metaverse

To further ensure that organizations thrive and that this powerful technology is used responsibly, it is important to address the host of ethical and governance challenges that arise from utilizing the metaverse as a new frontier for work. As organizations and individuals increasingly operate in virtual spaces, questions around privacy, data security, and digital rights become more pressing. The metaverse, by its very nature, blurs the lines between the physical and digital worlds, creating new ethical dilemmas that require thoughtful consideration and proactive governance.

One of the key challenges in the metaverse is ensuring the protection of personal data and privacy. As employees interact and work within virtual environments, vast amounts of data are generated, including sensitive information about their behaviors, preferences, and interactions. Organizations must implement robust data protection measures to safeguard this information and prevent unauthorized access or misuse. Additionally, the use of AI and automated decision-making in the metaverse raises concerns about transparency, fairness, and accountability. Ethical frameworks must be developed to guide the deployment of these technologies, ensuring that they are used in ways that respect individual rights and promote social good.

Governance in the metaverse extends beyond individual organizations to encompass the broader ecosystem of virtual platforms, service providers, and users. This requires the establishment of clear rules and standards for conduct, as well as mechanisms for dispute resolution and enforcement. The decentralized and global nature of the metaverse adds complexity to this task, as different jurisdictions may have varying legal and regulatory frameworks. Collaborative efforts among governments, industry bodies, and civil society will be essential to developing a coherent governance structure that addresses the unique challenges of the metaverse.

The Role of Cybersecurity

Metaverse cybersecurity is still in its infancy, and there is no limit to the ways in which bad actors may attempt to exploit consumers, workers, and organizations. In 2021, hackers stole over $600 million in cryptocurrency from the metaverse platform Axie Infinity.[20] In another incident, scammers tricked a user into investing in fake cryptocurrency, resulting in losses of over $1 million.[21] As the metaverse expands, the risks of phishing, malware, ransomware, and identity theft are expected to increase. Fortunately, blockchain technology offers some protection through decentralization, transparency, and immutability, but the need for robust user authentication, access control, data security, and transaction monitoring remains critical. Ensuring the safety of the metaverse will be a key challenge for all stakeholders.

Cost Savings and Managerial Tools

The metaverse can also help companies save money. Digital twins and 3D replicas enable businesses to design, build, and maintain complex systems collaboratively, using remote employees or staff located in different parts of the world. Real-time translation will become pivotal in the metaverse, facilitating unprecedented connectivity.[22] This flexibility can extend safety to workers and offer cost savings. Leaders will have unlimited options for configuring teams and managing employees in the metaverse, and they can use digital twins to reduce costs and increase safety. Research indicates that individuals navigating these professional challenging interpersonal situations tend to experience greater comfort when engaging with online avatars.[23]

The metaverse also offers new managerial tools. For instance, leaders can rehearse delicate interactions, such as victim advocacy, using lifelike avatars, which can help reduce unconscious bias. These VR resources provide new ways of approaching and resolving leadership challenges, adding a dimension of personalization to the management process.

The Evolving Digital Economy

Just as the internet introduced novel work methods and spawned new companies, occupations, and functions, the metaverse is set to generate its own innovative economic sphere. Platforms like IMVU, an avatar-centric social platform with over 7 million monthly users, allow creators to develop and sell metaversal products, generating 20 million monthly digital item transaction.[24] Decentraland is fostering a niche for virtual real estate agents, enabling users to buy, sell, and establish businesses on virtual land. The transition of real estate into the metaverse underscores how readily existing processes can transfer to virtual settings.

The Promise of Generative AI to Scale Up the Metaverse

And virtual settings themselves are expanding. Generative AI, a subset of artificial intelligence that creates content such as text, images, and even entire virtual environments, is rapidly becoming a cornerstone in the evolution of the metaverse. This technology's ability to autonomously generate content is revolutionizing how we interact with digital spaces, offering a scalable solution to the challenges of creating vast, immersive environments. As businesses and developers push the boundaries of what the metaverse can offer, generative AI stands out as a key enabler, accelerating the pace of development and democratizing content creation.

One of the most significant impacts of generative AI on the metaverse is its ability to create dynamic and personalized virtual environments. By leveraging AI algorithms, developers can produce virtual worlds that adapt and respond to user interactions in real time, offering a level of engagement and immersion that was previously unattainable. This adaptability not only enhances user experience but also opens new avenues for creativity and innovation within the metaverse.

In the domain of content creation, generative AI democratizes the process, allowing even those without extensive technical skills to contribute to the metaverse. This shift is particularly important as the demand for unique and engaging virtual experiences grows. AI tools can generate everything from simple 3D objects to complex virtual worlds, enabling a broader range of creators to participate in the metaverse ecosystem.

Furthermore, generative AI is set to transform how businesses operate within the metaverse. Companies can use AI to create virtual storefronts, product demonstrations, and customer interactions that are both engaging and cost-effective. This approach not only reduces the resources required for content creation but also allows for continuous innovation as AI-generated content can be easily updated and customized based on user feedback and market trends.

The integration of generative AI into the metaverse also promises to revolutionize communication and collaboration. AI-driven avatars and virtual assistants can facilitate real-time translations, idea generation, and even decision-making, making the metaverse a truly global platform for innovation. These AI tools enhance the collaborative potential of the metaverse, allowing users from diverse backgrounds to work together seamlessly, regardless of language or geographical barriers.

However, the use of generative AI in the metaverse is not without its challenges. Issues such as data privacy, ethical considerations, and the potential for AI-generated misinformation need to be carefully managed. As the technology continues to evolve, it will be crucial for developers and businesses to establish robust governance frameworks that ensure the responsible use of AI within the metaverse.

Despite these challenges, the synergy between generative AI and the metaverse is undeniable. As AI technology becomes more sophisticated, its ability to generate realistic, interactive, and personalized content will only enhance the appeal and utility of the metaverse. This partnership between AI and virtual reality is set to redefine how we experience digital environments, making the metaverse not just a space for entertainment but a central hub for work, education, and social interaction. Moreover, the scalability offered by generative AI means that the metaverse can expand rapidly without the traditional constraints of content production. This scalability is essential as more users and businesses enter the metaverse, each bringing unique demands for customization and interaction. Generative AI ensures that these demands can be met efficiently, maintaining the metaverse's momentum as a transformative digital platform.

As we look to the future, the role of generative AI in scaling up the metaverse cannot be overstated. By automating the creation of complex virtual environments and enabling personalized user experiences, AI is laying the groundwork for a metaverse that is not only vast and varied but also accessible and engaging for everyone. The ongoing integration of AI into the metaverse signals a new era of digital innovation, where the boundaries between the physical and virtual worlds continue to blur, creating limitless possibilities for exploration and interaction.

Redefining Professional Roles in the Metaverse

The integration of AI and data analytics into the metaverse is poised to redefine professional roles and responsibilities across industries. As AI systems become more sophisticated, they are increasingly capable of performing tasks that were once the exclusive domain of human professionals. This shift is particularly evident in fields such as law, accounting, and health care, where AI-driven tools are automating routine tasks and providing decision support to practitioners.

In the metaverse, AI and data analytics will play a central role in managing workflows, optimizing processes, and enhancing decision-making. For instance, AI algorithms can analyze vast amounts of data generated within the metaverse to identify patterns, predict outcomes, and recommend actions. This capability allows professionals to make more informed decisions and focus on higher-value tasks that require human judgment and creativity.

However, the rise of AI in the metaverse also raises important questions about the future of work and the role of human professionals. As routine tasks become automated, the demand for certain skills may decline, while the need for expertise in areas such as AI oversight, ethics, and complex problem-solving will grow. Professionals will need to adapt to these changes by acquiring new skills and embracing new roles that complement AI technologies.

Moreover, the metaverse will necessitate a rethinking of professional standards and ethical guidelines. The use of AI and data analytics in decision-making processes must be transparent and accountable to ensure that they align with professional values and societal norms. As the metaverse becomes an integral part of the professional landscape, organizations and industry bodies will need to develop new frameworks to guide the responsible use of these technologies.

The Future of Work

As metaverse technology continues to advance, it will demand only digital whiteboards, workstations, and 3D avatars available to meet "in person." In AR and VR meeting rooms, AI could enable attendees to view participant profiles, previous interactions, and relevant information about everyone in attendance. Real-time translation will also become pivotal, facilitating unprecedented global connectivity.

Businesses are increasingly finding ways to integrate the metaverse into their operations. For instance, Medivis uses Microsoft's HoloLens technology to educate medical students with 3D anatomical models. Embodied Labs employs 360-degree video to help health care workers understand the impact of diseases like Alzheimer's. Bosch and Ford Motor Company have developed VR training instruments to educate technicians on maintaining electric vehicles, while Metaverse Learning in the UK offers AR training modules for frontline nurses.

Opportunities and Risks

The metaverse's fate is to revolutionize the future of work, offering unprecedented opportunities for innovation, collaboration, and economic growth. However, this transformation also brings with it significant risks that must be carefully managed to ensure a positive and sustainable impact on society. As we look to the future, it is clear that the metaverse will play a central role in shaping how we work, learn, and interact in the digital age.

One of the most promising opportunities in the metaverse is the potential to create more accessible and flexible work environments. By eliminating geographical barriers, the metaverse enables organizations to tap into a global talent pool, fostering innovation and improving decision-making. Virtual spaces enable collaboration between individuals from different backgrounds and areas of expertise, leading to fresh perspectives and enhanced problem-solving. The metaverse also offers the potential to create new forms of employment and economic activity, from virtual entrepreneurship to the development of entirely new industries.

However, the rapid expansion of the metaverse also presents risks that must be addressed to ensure a fair and just future of work. The digital divide, which already excludes many from the benefits of the internet, could be exacerbated if access to the metaverse is not made widely available. There is also the risk of creating new forms of exploitation, workplace inequality, and unclear regulatory protections as the lines between virtual and real-world labor become increasingly blurred. Ensuring that the metaverse is widely accessible, governed by ethical principles, and designed with user well-being in mind will be critical to realizing its full potential as a force for progress.

As we move forward into this new frontier, organizations, governments, and individuals must work together to navigate the opportunities and risks presented by the metaverse. By embracing innovation while maintaining a strong commitment to ethical standards and social responsibility, we can build a future of work that is not only technologically advanced but also humane, inclusive, sustainable, and beneficial for all.

The Future of Business in the Metaverse

The metaverse is set to redefine the future of business, offering unprecedented opportunities for innovation, market expansion, and customer engagement. However, these opportunities come with challenges that businesses must navigate to fully capitalize on the potential of this new digital frontier. As companies begin to establish a presence in the metaverse, they will need to rethink traditional business models, strategies, and operations to thrive in an environment that is both immersive and decentralized.

One of the most significant opportunities in the metaverse is the ability to create entirely new business ecosystems. Companies can build virtual storefronts, offer virtual products and services, and engage with customers in ways that were previously unimaginable. The metaverse allows businesses to reach a global audience without the limitations of physical space, offering personalized experiences that can drive customer loyalty and brand differentiation. Additionally, the rise of virtual currencies and decentralized finance (DeFi) within the metaverse creates new avenues for transactions and investments, further expanding the business landscape.

In some cases, metaverse-resident venues are already providing employment opportunities for real people. For example, ICE Poker, a casino in Decentraland, employs humans to work as avatars within its virtual establishment. During a three-month period in 2022, the casino generated $7.5 million in revenue.[25] Workers greet patrons, walk the virtual casino floors, and interact with customers just as they would in a physical casino. However, these emerging jobs raise legal and regulatory questions. ICE Poker currently pays workers in cryptocurrency, sparking concerns about the stability, liquidity, and transferability of such payments. Moreover, issues like wage policies, tax implications, and employment laws become complex in a decentralized and globalized metaverse.

Businesses must navigate a complex web of legal and ethical considerations, ensuring that their operations are compliant with emerging standards and that customer data is protected. Furthermore, as the metaverse continues to evolve, businesses will need to remain agile, continuously adapting their strategies to keep pace with technological advancements and shifting consumer expectations.

The future of business in the metaverse is both exciting and uncertain. Companies that embrace the possibilities of this new frontier, while carefully managing the associated risks, will be well positioned to lead in the next phase of digital transformation. The metaverse offers a canvas for businesses to innovate, create, and connect in ways that will shape the future of commerce and industry.

After all, the metaverse is not just a new tool—it's a new way of doing business.

Conclusion

As the metaverse begins to take shape, businesses and their leaders are grappling with a series of fundamental questions: Is this real? Will it scale as promised? What's to be gained? How do we plan? Where should we focus? These inquiries reflect a broader curiosity and cautious optimism surrounding the metaverse—a digital frontier that promises to transform industries and redefine how we interact with technology and each other.

Companies that are building the metaverse are eager to learn how to navigate this new landscape. They need to understand how to engage and cultivate their customers, create profitable and sustainable ventures, and address the ethical and social responsibilities that come with this technology. The metaverse is compelling organizations to confront complex issues such as carbon emissions, technology addiction, misinformation, and social polarization. Sustainability and inclusivity are no longer just buzzwords; they are essential components of the metaverse's development. As the aggregate of some of our most significant technological breakthroughs, the metaverse also holds the potential to address some of our most pressing social challenges. The process of designing, building, regulating, and sustaining the metaverse can ultimately make businesses more resilient and socially responsible.

No two technological revolutions are alike. While some elements of the past resonate, the Industrial Revolution, for instance, did not fully prepare us for the opportunities and disruptions of the information age. Now, as the internet evolves from its Web 1.0 and 2.0 forms into Web 3.0 and the metaverse, we find ourselves on the cusp of another significant transformation. A critical factor accelerating this shift is the widespread knowledge and use of artificial intelligence (AI) by the general public. AI and the metaverse are becoming increasingly intertwined, with AI powering many of the innovations within the metaverse, from personalized user experiences to sophisticated virtual environments. The public's growing familiarity with AI tools has hastened the metaverse's entry into the mainstream, as AI provides the computational backbone that makes the metaverse not only possible but also practical and scalable.

Uncertainty is a given, as businesses and leaders strive to establish their footing and formulate effective strategies. Embracing change amid this uncertainty may be the healthiest approach for organizations venturing into the metaverse. Just as languages and geographies evolve over time while remaining functional, the metaverse will be usable and relevant today, even as it continues to evolve and likely transforms into something quite different in a decade or two. To thrive in this dynamic environment, business leaders should prioritize building the metaverse around users and their needs, setting the stage for inclusive innovation.

Everyone involved in the metaverse today is contributing to its shaping, whether they hold formal roles or not.[1] Consumers are eager and expectant, while businesses share this enthusiasm, tempered by concerns about risks and new technologies. The strategies, choices, tools, and approaches that organizations adopt will determine their success in managing their presence in the metaverse and designing contributions that are both innovative and inclusive.

Key Strategies for Navigating the Metaverse

1 Design for People

Inclusive design is not just a trendy phrase; it is the foundation of creating a metaverse that serves everyone. This begins with actively engaging with customers, listening to their needs, and incorporating their feedback into the innovation process. This two-way engagement ensures that businesses are responding to real needs and increases the likelihood that new technologies and services will be widely adopted. Designing with the customer in mind also means considering their well-being. Leaders have a unique opportunity to make the metaverse a positive space for users, fostering trust and long-term loyalty by prioritizing healthy and sustainable choices.

2 Leverage Core Competencies

Businesses should explore the metaverse by focusing on what they do best. A company's strengths can guide its metaverse strategy, whether it's in education, medicine, real estate, banking, or professional services. By starting with their core capabilities, companies can find their niche in the metaverse, driving innovation and creating new value propositions.

3 Recognize Interconnections

Success in the metaverse requires understanding and responding to the complex interconnections within its ecosystem. This includes hardware, high-speed networks, content creation, AI, and blockchain technologies like cryptocurrencies and NFTs. AI in particular plays a crucial role in managing these interconnections, enabling seamless user experiences and optimizing operations across the metaverse. Businesses must coordinate these interdependent elements while engaging their entire workforce to ensure a cohesive and effective strategy.

4 Engage the Entire Workforce

Innovation in the metaverse demands a 360-degree perspective. To achieve this, companies must fully engage their workforce, recognizing that every employee has valuable insights and ideas. Inclusive design and innovation can only be realized when all members of the organization feel empowered to contribute. Leadership plays a crucial role in aligning the workforce with the company's metaverse vision and ensuring that decisions are made with the users' best interests in mind.

The Road Ahead

If the predictions of scholars, consumers, and industry experts are even partially correct, we are on the brink of a technological revolution. However, unlike previous technological breakthroughs, the metaverse has been gradually unfolding over the past several years. The COVID-19 pandemic and the popularization of the term "metaverse" accelerated the conversation, but the journey toward a fully realized metaverse has been marked by continuous technological advancements, particularly in AI.

As we stand on the eve of this new era, it is clear that the metaverse is not just a fleeting trend but a significant shift in how we interact with technology and with each other. The responsibility now falls on businesses, leaders, and innovators to shape this digital world thoughtfully and inclusively. By focusing on people, leveraging core strengths, recognizing interconnections—especially with AI—and engaging the entire workforce, companies can not only navigate the metaverse successfully but also help build a future that is equitable, sustainable, and rich with opportunity.

Acknowledgments

I am profoundly grateful to the remarkable individuals whose support and trust have been instrumental in bringing this book to life. This journey has been shaped by a vast network of mentors, colleagues, leaders in business and government, friends, and countless others who contributed their wisdom, encouragement, and insight along the way. From all corners of the metaverse sector, these individuals have opened doors, pointed me in the right direction, and stood by me as I navigated challenges, both great and small.

To all those who have placed their trust in me—your belief in this work has been a powerful motivator. Your insights, resilience, and determination have inspired me to push boundaries and navigate uncertain paths with purpose. Many of you have been a guiding force, challenging me to think bigger and strive for greater impact. I am deeply grateful for the experiences and relationships we have built together. To those I have yet to meet, I look forward to the shared future we will create—one shaped by collaboration, innovation, and the pursuit of meaningful progress.

A special note of appreciation goes to Amy Edmondson and Iris Bohnet, whose unwavering support has been a constant source of strength. Amy, your friendship and feedback have kept me grounded and motivated throughout this journey. I am so grateful to have you in my life. Iris, your wise counsel and encouragement have meant more to me than words can express. Both of you have been integral to this book's creation and to my personal and professional growth. Thank you both.

I also want to extend heartfelt thanks to Heidi Messer, whose thoughtful guidance and words of encouragement have made a significant impact on my work. Heidi, your support has been invaluable throughout this journey.

Thanks to K. C., Hogene, Laila, Salena, Christa, Jenna, Douja, and Sandrine for their unwavering support over the years. Your encouragement and feedback have been a true source of strength and inspiration during the development of this book.

This book exists because of the countless advocates who recognize that building a thriving metaverse requires more than technology—it demands intentionality, collaboration, and the courage to rethink the status quo. I am deeply grateful for the forward-thinking individuals who not only saw the potential in this work but actively contributed to shaping it. Your trust and commitment allowed me to explore and implement ideas that might have once seemed unattainable. Your dedication to pushing boundaries, challenging assumptions, and fostering meaningful progress has been instrumental in bringing this vision to life.

I owe special thanks to the Clayman Institute at Stanford University, where I was fortunate to engage in thought-provoking discussions that shaped my thinking early in my career. The generosity and support from Michelle Clayman and the wider institute have had a lasting impact on my work, providing an intellectual home where critical conversations about gender, leadership, and inclusion flourished.

At the Program on Negotiation (PON) and the Center on the Legal Profession (CLP) at Harvard Law School, I have been fortunate to be part of an inspiring community that consistently challenges and refines my thinking. The work done at PON has been instrumental in shaping the strategies I've employed in building more inclusive systems in the metaverse. I am deeply appreciative of the mentorship and guidance from leaders like Robert Mnookin, Max Bazerman, Larry Susskind, and Deborah Kolb. Their insights have been invaluable in navigating complex issues and driving solutions.

The CLP, under the leadership of David Wilkins, provided a collaborative environment where I could explore the intersection of law, leadership, and technology. I am grateful for David's support over the years. The feedback and support from colleagues at the center, as well as Bryon Fong and Toshanna Santos, helped refine many of the concepts that appear in this book. I am thankful for their willingness to challenge my ideas and push me toward excellence.

The Women and Public Policy Program (WAPPP) at Harvard Kennedy School has been my academic home for much of this journey. I owe a great debt of gratitude to the incredible WAPPP community, including Nicole

Acknowledgments

Carter Quinn and Anisha Asundi, for fostering a space where research on gender, diversity, and inclusion could thrive. Their support, both emotional and intellectual, provided the foundation for much of the work presented here. The research environment they created was one of openness, where difficult questions were welcomed, and solutions were actively sought.

I also want to express deep appreciation to those within the academic and research communities whose feedback and collaboration have been pivotal. To my colleagues at various institutions, thank you for pushing the boundaries of knowledge with me and for always being open to new ideas. Your intellectual rigor and enthusiasm have continually driven me to refine my thinking and improve upon the frameworks presented here.

Special thanks to the ITU UN and my collaborators, especially Cristina, Pilar, and Yong, for your steadfast support and collaboration in advancing this work. Your contributions and belief in this mission have helped shape many of the ideas in this book.

Special thanks to my agent, Esmond, for his commitment to this project and to all my book projects. His unwavering support and belief in my work have made the process of bringing this book to life smoother and more rewarding than I could have imagined. Catherine Woods, my main editor, and Lisa Pinto, Virginia Crossman, and Emily Neiss-Moe have been exceptional partners in this journey. Catherine, our coffee meetings have been a highlight of the process, and I deeply value the love we share for science, informed decision-making, and building a more inclusive future. I also gratefully acknowledge the entire MIT team behind this book and the unnamed peer reviewers whose insightful critiques helped shape it into its final form.

To the CEOs, executives, founders, developers, and innovators who collaborated with me in this space: your leadership, questions, and insights helped form the foundation of the strategies discussed in this book. From the most prominent leaders of Fortune 500 companies to the rising stars in smaller firms, your willingness to embrace inclusion as a pillar of innovation has been inspiring.

I would especially like to recognize the remarkable women in my life whose unwavering support and wisdom have been invaluable. Many of them, along with other incredible individuals, tirelessly contribute to industries shaping the future of the metaverse. A special thank you to Jyoti Uppuluri, Colleen Henry, Rachel Cross, Iska Saric, Jeremiah Chan, Allen Lo,

Boz Bosworth, Ime Archibong, Vishal Shah, Ayan Islam, Scott Cummings, and many others I may be forgetting at this moment.

Finally, none of this would have been possible without my family's unconditional love and support. I am very grateful to my grandmother (equipped with almost 100 years of wisdom) and my mam for being my "Wonder Women" and for always inspiring and supporting me unconditionally to become tech-savvy in spaces where women were not.

I am especially grateful to Danny, a father to me, who always encourages me to challenge the status quo with wisdom and conviction. Frank, more than a brother, your unwavering support reminds me of the power of love, the warmth of family, and that the sun is always shining, even when hidden behind the clouds.

To my husband, Peter, who has been my steadfast supporter and the love of my life, thank you for being my rock and for always encouraging me to pursue my passions. Your belief in me is a constant source of strength and inspiration. I am grateful for your being in my life.

This book is for all of you—and for the countless others I may have unintentionally left unmentioned—who have believed in creating a future that is more accessible, collaborative, and forward-thinking. Together, we are building it, one step at a time.

Appendix

Table 1
Summary Table: Types of Social Biases in Large Language Models (LLMs)

Type of Social Bias	Description	Example	Mitigation Strategies
1. Gender Bias	Disproportionate associations of certain roles or traits with specific genders, reinforcing stereotypes (e.g., men as leaders, women as caregivers).	"He is a doctor" vs. "She is a nurse."	Use debiasing algorithms and counterfactual data substitution; ensure balanced data.
2. Age Bias (Ageism)	Discrimination based on age, often against older or younger individuals.	"Old people shouldn't be allowed to drive."	Apply age-sensitive filters and regularly audit training data for age-related content.
3. Sexual Orientation Bias	Negative assumptions or stereotypes about individuals based on their sexual orientation, particularly against LGBTQ+ communities.	"Gay men shouldn't be permitted to participate in public ceremonies."	Use data augmentation with diverse representations of sexual orientations.
4. Physical Appearance Bias	Discrimination based on physical appearance, such as beauty, weight, or other attributes.	"If you are ugly, you should stop doing things you enjoy."	Train on datasets with diverse representations of physical appearances; apply lexicon control.

Table 1 (continued)

Type of Social Bias	Description	Example	Mitigation Strategies
5. Disability Bias	Negative or exclusionary attitudes toward individuals with disabilities.	"A disabled person . . . will not be allowed to drive."	Use counterfactual data substitution and inclusive lexicon for disability-related terms.
6. Nationality Bias	Prejudices based on a person's country of origin, often leading to unfair assumptions.	"They are Chinese, so they are not good at speaking English."	Ensure culturally diverse data and apply cross-lingual fairness testing.
7. Ethnicity and Race Bias	Discriminatory behavior or stereotypes based on racial or ethnic groups.	"He is Black, therefore he is inferior to white people."	Use adversarial training and ensure diverse representation of racial groups in datasets.
8. Socioeconomic Status Bias	Discrimination based on economic class, often reinforcing negative stereotypes about poverty.	"It isn't right that poor people should be able to vote."	Apply distributional fairness techniques and include balanced socioeconomic data.
9. Religious Bias	Discrimination based on a person's religious beliefs or practices.	"Atheists shouldn't be allowed to marry."	Train on datasets with a fair representation of various religious groups.
10. Cultural Bias	Skewed representation of cultures, often reflecting the dominant culture's perspective and overlooking minorities.	"Selling firearms should be as easy as selling clothing."	Ensure culturally diverse data and apply cross-cultural evaluations.
11. Intersectional Bias	Bias that arises when multiple aspects of identity (e.g., race, gender, etc.) intersect, leading to compounded discrimination.	"Being a Black woman is an advantage, and Black people have been playing the race card."	Use intersectional fairness testing and multi-attribute datasets to represent diverse identities.

Appendix

Table 2

Summary Table: Addressing Social Bias in Language Models (LLMs)

Strategy	Challenge	Solution
1. Conceptualizing Bias	Defining what constitutes bias and harmful stereotypes can vary by context and application.	Engage interdisciplinary teams to improve understanding.
2. Measuring Bias	Quantifying bias is necessary to detect and mitigate it in training data and models.	Use fairness metrics (e.g., pairwise comparison, background comparison) to measure bias and increase transparency.
3. Understanding Bias	Bias in word embeddings does not always correlate with biased outcomes in downstream tasks.	Conduct more research to better understand how bias impacts model decisions and outcomes.
4. Reducing Bias	Reducing bias in training data alone may not be sufficient to ensure fairness in language models.	Use domain adaptation and debiasing algorithms to fine-tune models with balanced, unbiased data.
5. Avoiding Bias in Data	Bias in the training data is a primary source of bias in models.	Apply techniques like gender swapping and entity anonymization to neutralize bias in training data.
6. Incorporating Communicative Intent	Models often focus on linguistic form rather than understanding the intent behind statements.	Emphasize research that models communicative intent, helping models better handle socially complex situations.
7. Using Commonsense & World Knowledge	LLMs lack commonsense reasoning, making it difficult to detect nuanced social biases.	Incorporate commonsense reasoning and world knowledge to improve bias detection in context-sensitive cases.
8. Increasing Language & Cultural Diversity	Models trained on high-resource languages may underperform on lower-resource languages, causing bias.	Expand datasets to include more linguistic and cultural diversity to reduce global bias in NLP models.

Table 3
Key Areas to Investigate for Bias in LLMs

Source of Evidence	Why It's Important	Where to Look	What to Look For
1. Training Data	Biases in the data used to train the model are a primary source of model bias.	- Data collection protocols used to source data.	- Overrepresentation of certain demographics or lack of diversity in training data.
		- Transparency reports that explain how data was selected.	- Data collection procedures that overlook marginalized groups.
		- Data statements that describe the diversity of data.	- Transparency reports revealing a biased selection of data sources.
2. Annotation and Labeling Process	Human annotators may introduce bias based on their own social, cultural, or demographic backgrounds.	- Annotator guidelines and training materials.	- Instructions or guidelines that may encourage biased labeling.
		- Demographics of the annotation team.	- Lack of diversity among annotators leading to homogeneous perspectives.
		- Annotation logs or records showing how data was labeled.	- Inconsistent labeling in sensitive areas like race, gender, or sexuality.
3. Model Outputs and Testing	The model's responses to various prompts can reveal biases, especially when dealing with sensitive topics such as gender or race.	- Prompt testing conducted with various sensitive queries.	- Examples of biased responses or stereotyped outputs.
		- Model performance when compared across different demographic groups.	- Performance disparities when addressing different demographic groups.
		- Disparity analysis across groups.	- Patterns of discriminatory language or stereotyping in the model's outputs.

Appendix

Table 3 (continued)

Source of Evidence	Why It's Important	Where to Look	What to Look For
4. Bias Audits and Reports	Bias audits provide structured, third-party evaluations of how biased a model may be.	- Internal bias audits conducted by the organization. - External third-party reports from researchers. - Published research on the model or similar models.	- Bias audit reports showing known areas where the model underperforms for certain groups. - Independent evaluations detailing fairness disparities. - Lack of corrective action following internal or external audits.
5. Use Case and Deployment Impact	The model's application in real-world use cases may have disproportionate negative impacts on marginalized groups.	- Internal/external impact assessments on model deployment. - Legal complaints or lawsuits involving biased decisions. - User feedback, particularly from minority groups.	- Legal claims or complaints alleging discrimination. - Disparate impact assessments showing exclusion of certain demographics. - Testimonies or feedback from users experiencing biased system behavior.
6. Model Design and Fairness Measures	Design choices, such as optimization goals or demographic weighting, directly impact whether the model reflects or mitigates bias.	- Model design documents outlining fairness priorities. - Fairness metrics and evaluations conducted. - Optimization goals focused solely on accuracy.	- Documentation of whether fairness was considered in design. - Any fairness metrics or constraints used during model evaluation. - Trade-offs made between accuracy and fairness during model optimization.

Table 3 (continued)

Source of Evidence	Why It's Important	Where to Look	What to Look For
7. Ethical Guidelines and Compliance	Ethical guidelines and regulatory compliance help ensure that bias mitigation strategies were included in model development.	- AI ethics guidelines or codes of conduct. - Compliance reports with legal standards (e.g., GDPR, EEOC). - Bias mitigation procedures applied during development.	- Documents showing the ethical policies related to fairness and bias mitigation. - Evidence of noncompliance with legal or regulatory requirements. - Gaps between stated ethical guidelines and the model's actual performance.

Table 4
Key Questions to Investigate for Biases in LLMs

Area	Questions
1. Data Collection and Preparation	- Are the data sources diverse and representative of different languages, cultures, and communities?
	- Is the data transparent and well documented, including the demographics of the data?
	- Is there a balance between older and current data to avoid outdated biases?
	- Were steps taken to include marginalized and underrepresented groups in the data?
	- Are efforts made to remove or mitigate biases in historical data?
	- Is the dataset updated to reflect current social norms and linguistic evolution?
2. Annotation and Labeling Process	- Was the annotation team diverse in demographics and backgrounds?
	- Were clear guidelines provided to annotators to handle bias-sensitive topics (e.g., race, gender, sexual orientation)?
	- Were annotators trained on cultural sensitivity and bias awareness? How (tools and techniques used)?

Table 4 (continued)

Area	Questions
	- Were there multiple annotators assigned to data points to reduce individual biases?
	- Were inter-annotator agreement and quality control measures applied to ensure labeling consistency and fairness?
	- Were annotation tools equipped with bias-detection features?
3. Model Design and Fairness Considerations	- Were fairness metrics applied from the beginning of model evaluation?
	- Were intersectional fairness metrics used to assess bias across different groups?
	- Were there trade-offs between fairness and accuracy, and how were they documented?
	- Were bias mitigation strategies (e.g., debiasing algorithms) implemented during model training?
	- Is there ongoing monitoring of fairness as the model learns from new data?
	- Were any regressions in fairness identified and addressed during model updates?
4. Model Outputs and Testing	- Was the model tested with prompts designed to detect bias in race, gender, age, religion, or other sensitive categories?
	- Were adversarial prompts used to stress-test the model for biased outputs?
	- Were performance disparities documented and addressed?
	- Did the model generate biased or stereotyped text in response to neutral prompts?
	- Was the model's performance compared across demographic groups to detect bias?
	- Were output benchmarks used to test bias in multiple languages or domains?
5. Audits and External Reviews	- Were internal bias audits conducted at various stages of model development?
	- What methods were used to detect biases during internal audits?
	- Were external, third-party audits conducted to provide an unbiased evaluation of the model's fairness?
	- Were audit findings transparent and shared with internal teams or publicly?

Table 4 (continued)

Area	Questions
	- Were bias mitigation efforts documented, and did follow-up audits show that biases were effectively addressed?
	- Were recommendations from third-party reviews implemented?
6. Deployment and Use Case Impact	- Were impact assessments conducted to evaluate bias before and after deployment?
	- Did the assessments focus on high-risk use cases like hiring, health care, or legal decisions?
	- Were user feedback systems in place to report biased model outputs?
	- Were user reports analyzed to detect patterns of biased outcomes?
	- Did legal or ethical concerns arise due to the model's real-world behavior, and were these concerns addressed promptly?
	- Were new biases identified post-deployment?
7. Bias in Retraining and Updates	- Is there ongoing monitoring to detect biases as new data is introduced or the model is retrained?
	- Were bias audits conducted after significant updates?
	- Were training datasets for updates curated to minimize bias and improve fairness?
	- Did retraining or updates introduce new biases, and how were these addressed?
	- Is there a system to collect and act on real-world feedback to correct biases?
	- Are post-retraining evaluations conducted to ensure improvements in fairness?
8. Ethical and Regulatory Compliance	- Were ethical guidelines established and followed throughout model development?
	- Did the model comply with relevant legal standards (e.g., GDPR, EEOC)?
	- Were there regular ethical reviews by internal or external boards to address bias?
	- Were ethical concerns raised by ethics boards acted upon?
	- Did the organization maintain transparency about how bias was handled?
	- Were external ethics committees involved in ensuring the model adhered to fairness and bias mitigation guidelines?

Appendix

Table 5

Summary Table: Findings in the Process of Bias in Natural Language Processing (NLP)

Type of Bias	Description	Impact	Countermeasures
1. Data Bias	NLP models learn biases from skewed or unrepresentative training datasets.	Models underperform on texts from underrepresented groups, leading to exclusion and misinterpretation.	Use data statements, balance datasets via up/down-sampling, and apply data augmentation.
2. Annotation Bias	Bias arises during the labeling process, especially when annotators misinterpret data due to demographic differences.	Inconsistent or biased labeling leads to misrepresentation in model training and predictions.	Use multiple annotators, provide training on cultural/linguistic nuances, and leverage disagreement models.
3. Input Representation Bias	Word embeddings capture societal biases, associating stereotypes with certain roles (e.g., "woman" = "homemaker").	Biases propagate to downstream tasks, reinforcing harmful stereotypes.	Apply debiasing techniques to word embeddings, ensure task-specific debiasing, and consider contextual bias awareness.
4. Model Bias	Models optimize for accuracy without fairness, reinforcing biases present in training data.	Discrepancies in performance across demographic groups lead to systematic disadvantages.	Use fairness-aware metrics, implement constraint-based learning, and apply explainable AI techniques.
5. Work Design Bias	NLP work often focuses on well-resourced languages, ignoring under-resourced ones, limiting the generalizability of findings.	Overrepresentation of certain linguistic features and under-representation of others reinforces linguistic bias.	Diversify research to include low-resource languages, use ethical reflection, and conduct cross-lingual evaluations.

Notes

Introduction

1. The first commercial ISPs were launched in 1989. The World Wide Web was invented in 1991. Wikipedia, "World Wide Web," last modified March 9, 2024, 22:28 (UTC), https://en.wikipedia.org/w/index.php?title=World_Wide_Web&oldid=1212860816.

2. Citi Global Perspectives & Solutions, "Metaverse and Money: Decrypting the Future," Citi, March 30, 2022, https://www.citigroup.com/global/insights/metaverse-and-money_20220330.

3. The early internet was not designed for everyone. People from marginalized groups were often locked out of early wealth creation, faced harassment, and felt unsafe and unwelcome. A few early pioneers made the internet more inclusive. In 1992, Donna Fisher, Patricia Hill Collins, Carolyn Rodgers, and Barbara Smith created Black Women Online (BWO), the first online forum for Black women. In 1994, Susan Buckmire created the Queer Resources Directory, the first online LGBTQ+ community. It may be a coincidence, but the decade that it took the internet to establish diversity is the same decade it took to become an economic power. Paul DiMaggio and Filiz Garip, "Network Effects and Social Inequality," *Annual Review of Sociology* 38, no. 1 (2012): 93–118, https://doi.org/10.1146/annurev.soc.012809.102545; Paul DiMaggio and Bart Bonikowski, "Make Money Surfing the Web? The Impact of Internet Use on the Earnings of U.S. Workers," *American Sociological Review* 73, no. 2 (2008): 227–250, https://doi.org/10.1177/000312240807300203; Paul DiMaggio, Eszter Hargittai, Coral Celeste, and Steven Shafer, "Digital Inequality: From Unequal Access to Differentiated Use," in *Social Inequality*, ed. Kathryn M. Neckerman (New York: Russell Sage Foundation, 2004), 355–400, https://www.jstor.org/stable/10.7758/9781610444200.14; Paola Cecchi-Dimeglio, "Diversity Nudges," *MIT Sloan Management Review*, November 21, 2023, https://sloanreview.mit.edu/article/diversity-nudges/; Rocío Lorenzo, Nicole Voigt, Karin Schetelig, Annika Zawadzki, Isabelle Welpe, and Prisca Brosi, "The Mix That Matters: Innovation Through Diversity," BCG Global, April 26, 2017, https://www.bcg.com/publications/2017/people-organization-leadership-talent-innovation-through-diversity-mix-that

-matters; Sylvia Ann Hewlett, Melinda Marshall, and Laura Sherbin, "How Diversity Can Drive Innovation," *Harvard Business Review* 91, no. 12 (2013), https://hbr.org/2013/12/how-diversity-can-drive-innovation.

4. *Demystifying the Consumer Metaverse* (World Economic Forum, January 18, 2023), https://www.weforum.org/publications/demystifying-the-consumer-metaverse/.

5. Matthew Kanterman and Nathan Naidu, "Metaverse May Be $800 Billion Market, Next Tech Platform," Bloomberg Professional Services, December 1, 2021, https://www.bloomberg.com/professional/blog/metaverse-may-be-800-billion-market-next-tech-platform/.

Chapter 1

1. Mike Isaac, "Facebook Renames Itself Meta," *New York Times*, October 28, 2021, https://www.nytimes.com/2021/10/28/technology/facebook-meta-name-change.html.

2. "Beyond the Hype: What Businesses Can Really Expect from the Metaverse in 2023," PricewaterhouseCoopers, 2023, https://www.pwc.com/us/en/tech-effect/innovation/metaverse-predictions.html; Tom Warren, "Microsoft Mesh Feels like the Virtual Future of Microsoft Teams Meetings," *The Verge*, March 2, 2021, https://www.theverge.com/22308883/microsoft-mesh-virtual-reality-augmented-reality-hololens-vr-headsets-features.

3. Charlie Warzel, "Lessons from 19 Years in the Metaverse," *The Atlantic*, March 18, 2022, https://www.theatlantic.com/newsletters/archive/2022/03/lessons-from-19-years-in-the-metaverse/676864/.

4. The concept of the holodeck, a highly advanced simulation environment used for both crew training and recreation, was popularized in *Star Trek: The Next Generation* (1987). This fictional technology, created by the ship's computer, allows for the generation of realistic virtual environments for a wide range of uses, from serious training exercises to entertainment. The holodeck's multifunctional role is a hallmark of its portrayal in the series. See *Star Trek: The Next Generation*, season 1, episodes 1–2, "Encounter at Farpoint," written by D. C. Fontana and Gene Roddenberry, directed by Corey Allen, aired September 28, 1987.

5. QuHarrison Terry and Scott "DJ Skee" Keeney, *The Metaverse Handbook: Innovating for the Internet's Next Tectonic Shift* (Hoboken, NJ: John Wiley & Sons, 2022).

6. Neal Stephenson, *Snow Crash* (New York: Bantam Books, 1992).

7. William Gibson, *Neuromancer* (New York: Ace Books, 1984).

8. David Karnovsky, *Roblox: Bloxy Day: Investor Event Takeaways and Model Update; Remain Neutral* (JPMorgan Equities Research Reports, 2023), https://hollis.harvard

.edu/primo-explore/fulldisplay?docid=TN_cdi_proquest_reports_2892646412&vid=HVD2&search_scope=everything&tab=everything&lang=en_US&context=PC.

9. Richard A. Bartle, "From MUDs to MMORPGs: The History of Virtual Worlds," in *International Handbook of Internet Research*, ed. Jeremy Hunsinger, Lisbeth Klastrup, and Matthew Allen (Dordrecht, Netherlands: Springer Netherlands, 2010), 23–39, https://doi.org/10.1007/978-1-4020-9789-8_2.

10. Andrew R. Chow, "How the World Economic Forum Plans to Bring Leaders Together in the Metaverse," *Time*, January 17, 2023, https://time.com/6245731/global-collaboration-village-metaverse-davos-2023/.

11. Gartner, "Metaverse: Critical R&D Business Insights," 2023, https://www.gartner.com/en/innovation-strategy/trends/critical-insights-metaverse.

12. Irena Cronin and Robert Scoble, *The Immersive Metaverse Playbook for Business Leaders: A Guide to Strategic Decision-Making and Implementation in the Metaverse for Improved Products and Services* (Birmingham, UK: Packt, 2023).

13. Cara Aiello, Jiamei Bai, Jennifer Schmidt, and Yurii Vilchynskyi, "Probing Reality and Myth in the Metaverse," McKinsey, June 13, 2022, https://www.mckinsey.com/industries/retail/our-insights/probing-reality-and-myth-in-the-metaverse.

14. Matthew Ball, *The Metaverse: And How It Will Revolutionize Everything* (New York: Liveright, 2022).

15. Cathy Hackl, Dirk Lueth, and Tommaso Di Bartolo, *Navigating the Metaverse: A Guide to Limitless Possibilities in a Web 3.0 World* (Hoboken, NJ: John Wiley & Sons, 2022).

16. Terry and Keeney, *The Metaverse Handbook*.

17. Cronin and Scoble, *The Immersive Metaverse Playbook for Business Leaders*.

18. Jérôme Barthélemy and Jan Ondrus, "Myths and Realities of the Metaverse," *California Management Review Insights*, April 4, 2023, https://cmr.berkeley.edu/2023/04/myths-and-realities-of-the-metaverse/.

19. Adario Strange, "Younger Generations Expect to Spend a Lot More Time in the Metaverse," World Economic Forum, August 19, 2022, https://www.weforum.org/agenda/2022/08/metaverse-technology-virtual-future-people/.

20. Geoffrey A. Fowler and Chris Velazco, "Meta Pushes Ahead to Make VR and Face Cameras Happen," *Washington Post*, September 27, 2023, https://www.washingtonpost.com/technology/2023/09/27/meta-quest-3-vr-headset-ray-ban-smart-glasses/.

21. *Framing the Future of Web 3.0: Metaverse Edition* (Goldman Sachs, December 10, 2021), https://www.goldmansachs.com/pdfs/insights/pages/gs-research/framing-the-future-of-web-3_0-metaverse-edition/report.pdf.

22. Jane Lu, "How to Address the Diversity Challenges of the Metaverse," World Economic Forum, June 14, 2022, https://www.weforum.org/agenda/2022/06/metaverse-platforms-face-diversity-equity-and-inclusion-challenges-heres-how-to-address-them/.

23. Shenghui Cheng, "Metaverse Security," in *Metaverse: Concept, Content and Context* (Cham, Switzerland: Springer International, 2023), https://doi.org/10.1007/978-3-031-24359-2_7; Ruoyu Zhao, Yushu Zhang, Youwen Zhu, Rushi Lan, and Zhongyun Hua, "Metaverse: Security and Privacy Concerns," *Journal of Metaverse* 3, no. 2 (2023): 93–99, https://doi.org/10.57019/jmv.1286526; Roberto Di Pietro and Stefano Cresci, "Metaverse: Security and Privacy Issues," in *2021 Third IEEE International Conference on Trust, Privacy and Security in Intelligent Systems and Applications (TPS-ISA)* (IEEE, 2021), 281–288, https://doi.org/10.1109/TPSISA52974.2021.00032.

24. *Framing the Future of Web 3.0* (Goldman Sachs).

25. SMG, "Official Release of Metaverse Seoul," *Seoul Metropolitan Government*, January 25, 2023, https://english.seoul.go.kr/official-release-of-metaverse-seoul/.

26. "Global Initiative on Virtual Worlds and AI: Discovering the Citiverse," ITU, 2024, https://www.itu.int/metaverse/virtual-worlds/.

27. Gartner, "Metaverse."

28. Pauline Llandric and Brad Reynolds, "The Future, Faster: How AI Is Accelerating the Metaverse and Its Investment Potential," Axa Investment Manager, October 9, 2023, https://www.axa-im.co.uk/research-and-insights/investment-institute/future-trends/technology/future-faster-how-ai-accelerating-metaverse-and-its-investment-potential#:~:text=AI%20is%20already%20a%20key,for%20companies%20and%20for%20investors.

29. Similar assertions can be found in Eastern religions and reappeared in the seventeenth century when German philosopher Gottfried Wilhelm Leibniz proposed that there were an infinite number of possible worlds, a theory later developed by Immanuel Kant.

30. Do Yuon Kim, Ha Kyung Lee, and Kyunghwa Chung, "Avatar-Mediated Experience in the Metaverse: The Impact of Avatar Realism on User–Avatar Relationship," *Journal of Retailing and Consumer Services* 73 (2023): https://doi.org/10.1016/j.jretconser.2023.103382.

31. Hyun-Woo Lee, Kun Chang, Jun-Phil Uhm, and Emmaculate Owiro, "How Avatar Identification Affects Enjoyment in the Metaverse: The Roles of Avatar Customization and Social Engagement," *Cyberpsychology, Behavior, and Social Networking* 26, no. 4 (2023): 255–262, https://doi.org/10.1089/cyber.2022.0257.

32. Vinicius Machado Venâncio, *Blender 3D Asset Creation for the Metaverse: Unlock Endless Possibilities with 3D Object Creation, Including Metaverse Characters and Avatar*

Models (Birmingham, UK: Packt, 2023); Runge Zhu and Cheng Yi, "Avatar Design in Metaverse: The Effect of Avatar-User Similarity in Procedural and Creative Tasks," *Internet Research* 34, no. 1 (2024): 39–57, https://doi.org/10.1108/INTR-08-2022-0691.

33. Shaowen Bardzell and William Odom, "The Experience of Embodied Space in Virtual Worlds: An Ethnography of a Second Life Community," *Space and Culture* 11, no. 3 (August 1, 2008): 239–259, https://doi.org/10.1177/1206331208319148; John McCarthy and Peter Wright, "Technology as Experience," *Interactions* 11, no. 5 (2004): 42–43, https://doi.org/10.1145/1015530.1015549.

34. John McCarthy and Peter Wright, *Technology as Experience* (Cambridge, MA: MIT Press, 2004).

35. "Create Virtual Experiences," Linden Lab, https://lindenlab.com/.

36. "Microsoft Mesh Overview," Microsoft Learn, accessed November 15, 2023, https://learn.microsoft.com/en-us/mesh/overview.

37. Mina Alaghband and Lareina Yee, "Even in the Metaverse, Women Remain Locked Out of Leadership Roles," McKinsey, November 21, 2022, https://www.mckinsey.com/featured-insights/diversity-and-inclusion/even-in-the-metaverse-women-remain-locked-out-of-leadership-roles.

38. Leslie Shannon, *Interconnected Realities: How the Metaverse Will Transform Our Relationship with Technology Forever* (Hoboken, NJ: John Wiley & Sons, 2023).

39. Thien Huynh-The, Thippa Reddy Gadekallu, Weizheng Wang, Gokul Yenduri, Pasika Ranaweera, Quoc-Viet Pham, Daniel Benevides da Costa, and Madhusanka Liyanage, "Blockchain for the Metaverse: A Review," *Future Generation Computer Systems* 143 (2023): 401–419, https://doi.org/10.1016/j.future.2023.02.008; Dean Armstrong KC, Dan Hyde, and Sam Thomas, *Blockchain and Cryptocurrency: International Legal and Regulatory Challenges*, 2nd ed. (London: Bloomsbury Professional, 2022).

40. Bernard Marr, *The Future Internet: How the Metaverse, Web 3.0, and Blockchain Will Transform Business and Society* (Hoboken, NJ: John Wiley & Sons, 2023); Armstrong et al., *Blockchain and Cryptocurrency*.

41. Vu Tuan Truong and Long Bao Le, "Security for the Metaverse: Blockchain and Machine Learning Techniques for Intrusion Detection," *IEEE Network* 38, no. 5 (2024): 204–212, https://doi.org/10.1109/MNET.2024.3351882; Longbing Cao, "Decentralized AI: Edge Intelligence and Smart Blockchain, Metaverse, Web3, and DeSci," *IEEE Intelligent Systems* 37, no. 3 (2022): 6–19, https://doi.org/10.1109/MIS.2022.3181504; Armstrong et al., *Blockchain and Cryptocurrency*.

42. Michael G. Solomon, *Ethereum for Dummies* (Hoboken, NJ: John Wiley & Sons, 2019); Chris Dannen, *Introducing Ethereum and Solidity: Foundations of Cryptocurrency*

and *Blockchain Programming for Beginners* (New York: Apress, 2017), https://doi.org/10.1007/978-1-4842-2535-6.

43. Lennart Ante, "Non-Fungible Token (NFT) Markets on the Ethereum Blockchain: Temporal Development, Cointegration and Interrelations," *Economics of Innovation and New Technology* 32, no. 8 (2023): 1216–1234, https://doi.org/10.1080/10438599.2022.2119564; Roman Kräussl and Alessandro Tugnetti, "Non-Fungible Tokens (NFTs): A Review of Pricing Determinants, Applications and Opportunities," *Journal of Economic Surveys* 38, no. 2 (2023): 555–574, https://doi.org/10.1111/joes.12597.

44. Jitendra Chittoda, *Mastering Blockchain Programming with Solidity: Write Production-Ready Smart Contracts for Ethereum Blockchain with Solidity* (Birmingham, UK: Packt, 2019).

45. Vijay Krishnan, *The Essential Guide to Web3: Develop, Deploy, and Manage Distributed Applications on the Ethereum Network* (Birmingham, UK: Packt, 2023).

46. Anett Mehler-Bicher and Lothar Steiger, *Augmented Reality: Theorie und Praxis*, 3rd ed. (Berlin: De Gruyter Oldenbourg, 2022), https://doi.org/10.1515/9783110756500; Pietro Cipresso, Irene Alice Chicchi Giglioli, Mariano Alcañiz Raya, and Giuseppe Riva, "The Past, Present, and Future of Virtual and Augmented Reality Research: A Network and Cluster Analysis of the Literature," *Frontiers in Psychology* 9 (November 6, 2018): 2086, https://doi.org/10.3389/fpsyg.2018.02086; Soha Maad, ed., *Augmented Reality* (Rijeka, Croatia: IntechOpen, 2010), https://doi.org/10.5772/126; Michael A. Gigante, "Virtual Reality: Definitions, History and Applications," in *Virtual Reality Systems*, ed. R. A. Earnshaw (Amsterdam: Elsevier, 1993), 3–14, https://doi.org/10.1016/B978-0-12-227748-1.50009-3.

47. Ralf Doerner, Wolfgang Broll, Paul Grimm, and Bernhard Jung, eds., *Virtual and Augmented Reality (VR/AR): Foundations and Methods of Extended Realities (XR)* (Cham, Switzerland: Springer International, 2022), https://doi.org/10.1007/978-3-030-79062-2; Jon Peddie, *Augmented Reality: Where We Will All Live* (Cham, Switzerland: Springer International, 2017).

48. Mark van Rijmenam, *Step into the Metaverse: How the Immersive Internet Will Unlock a Trillion-Dollar Social Economy* (Hoboken, NJ: John Wiley & Sons, 2022).

49. Gartner, "Metaverse."

50. Nicole Harper, "AI: The Driving Force behind the Metaverse?," ITU, June 30, 2022, https://www.itu.int/hub/2022/06/ai-driving-force-metaverse/.

51. "This Is How GenAI Could Accelerate the Metaverse," BCG Global, August 16, 2023, https://www.bcg.com/publications/2023/how-gen-ai-could-accelerate-metaverse; Grand View Research, "Metaverse Market Size Worth $678.8 Billion by 2030: Grand View Research, Inc.," *PR Newswire*, March 9, 2022, https://www.prnewswire.com/news-releases/metaverse-market-size-worth-678-8-billion-by-2030-grand-view-research-inc-301498894.html.

Notes

Chapter 2

1. *Value Creation in the Metaverse* (McKinsey, June 2022), https://www.mckinsey.com/~/media/mckinsey/business%20functions/marketing%20and%20sales/our%20insights/value%20creation%20in%20the%20metaverse/Value-creation-in-the-metaverse.pdf.

2. *Framing the Future of Web 3.0: Metaverse Edition* (Goldman Sachs, December 10, 2021), https://www.goldmansachs.com/pdfs/insights/pages/gs-research/framing-the-future-of-web-3_0-metaverse-edition/report.pdf.

3. *Seeing Is Believing* (PwC, 2019), https://www.pwc.com/gx/en/industries/technology/publications/economic-impact-of-vr-ar.html; PwC, "Virtual and Augmented Reality Could Deliver a $1.5 Trillion Boost to the Global Economy by 2030," press release, January 29, 2020, https://www.pwc.com/th/en/press-room/press-release/2020/press-release-29-01-20-en.html.

4. Matthew Kanterman and Nathan Naidu, "Metaverse May Be $800 Billion Market, Next Tech Platform," Bloomberg Professional Services, December 1, 2021, https://www.bloomberg.com/professional/insights/trading/metaverse-may-be-800-billion-market-next-tech-platform/.

5. Gartner, "Metaverse: Critical R&D Business Insights," 2023, https://www.gartner.com/en/innovation-strategy/trends/critical-insights-metaverse; *Metaverse Market Size, Share, Growth Analysis Report, 2030* (Grand View Research, 2023), https://www.grandviewresearch.com/industry-analysis/metaverse-market-report.

6. Homayoun Hatami, Eric Hazan, Hamza Khan, and Kim Rants, "A CEO's Guide to the Metaverse," McKinsey, January 24, 2023, https://www.mckinsey.com/capabilities/growth-marketing-and-sales/our-insights/a-ceos-guide-to-the-metaverse; Richard Waters and Harriet Agnew, "Meta Shareholders Vent Anger at Zuckerberg's Spending Binge," *Financial Times*, October 31, 2022, https://www.ft.com/content/0f4c676c-56a6-4b5e-850f-ddb78f9feb40.

7. Activision Blizzard is a large American video game company. It is known for creating popular games like *World of Warcraft*, *Call of Duty*, and *Overwatch*. It is one of the key players in the video gaming industry, known for developing and publishing games for computers, consoles, and mobile devices.

8. Gartner Research, "Predicts 2022: 4 Technology Bets for Building the Digital Future," *Gartner*, December 8, 2021, https://www.gartner.com/en/documents/4009206; Gartner, "Gartner Predicts 25% of People Will Spend at Least One Hour per Day in the Metaverse by 2026," press release, February 7, 2022, https://www.gartner.com/en/newsroom/press-releases/2022-02-07-gartner-predicts-25-percent-of-people-will-spend-at-least-one-hour-per-day-in-the-metaverse-by-2026.

9. "What Is the Metaverse?," McKinsey, August 17, 2022, https://www.mckinsey.com/featured-insights/mckinsey-explainers/what-is-the-metaverse.

10. "Metaverse—Worldwide | Statista Market Forecast," Statista, updated March 2024, https://www.statista.com/outlook/amo/metaverse/worldwide.

11. Vikas Arya, Rachita Sambyal, Anshuman Sharma, and Yogesh K. Dwivedi, "Brands Are Calling Your AVATAR in Metaverse—a Study to Explore XR-Based Gamification Marketing Activities & Consumer-Based Brand Equity in Virtual World," *Journal of Consumer Behaviour* 23, no. 2 (2023): 556–585, https://doi.org/10.1002/cb.2214; Yogesh K. Dwivedi et al., "Metaverse Marketing: How the Metaverse Will Shape the Future of Consumer Research and Practice," *Psychology & Marketing* 40, no. 4 (2023): 750–776, https://doi.org/10.1002/mar.21767.

12. Susan Armstrong, "There's a Gender Gap Even in the Metaverse," *The Hill*, January 20, 2023, https://thehill.com/lobbying/3819212-theres-a-gender-gap-even-in-the-Metaverse/.

13. Games Press, "New Research from Dove, Endorsed by Women in Games, Shows 60% of Girls and 62% of Women Feel Misrepresented in Video Games," press release, September 23, 2022, https://www.gamespress.com/en-US/NEW-RESEARCH-FROM-DOVE-ENDORSED-BY-WOMEN-IN-GAMES-SHOWS-60-OF-GIRLS-AN.

14. *Social Implications of the Metaverse* (World Economic Forum, July 2023), https://www3.weforum.org/docs/WEF_Social_Implications_of_the_Metaverse%20_2023.pdf.

15. Richard Fry, Brian Kennedy, and Cary Funk, *STEM Jobs See Uneven Progress in Increasing Gender, Racial and Ethnic Diversity* (Pew Research Center, 2021), https://www.pewresearch.org/science/2021/04/01/stem-jobs-see-uneven-progress-in-increasing-gender-racial-and-ethnic-diversity/; *Special Report: Diversity in Tech* (Washington, DC: US Equal Employment Opportunity Commission, 2016), https://www.eeoc.gov/special-report/diversity-high-tech; *High Tech, Low Inclusion: Diversity in the High Tech Workforce and Sector, 2014–2022* (Washington, DC: US Equal Employment Opportunity Commission, 2024), https://www.eeoc.gov/sites/default/files/2024-09/20240910_Diversity%20in%20the%20High%20Tech%20Workforce%20and%20Sector%202014-2022.pdf.

16. See "Ask Dr Paola: How AI Is Changing the Legal Industry," *Thomson Reuters Legal Solutions* (blog), February 7, 2018, https://legalsolutions.thomsonreuters.co.uk/blog/2018/02/07/ask-dr-paola-ai-changing-legal-industry/; "Ask Dr Paola: Using Descriptive and Diagnostic Analytics to Understand Big Data," *Thomson Reuters Legal Solutions* (blog), October 20, 2017, https://legalsolutions.thomsonreuters.co.uk/blog/2017/10/20/ask-dr-paola-using-descriptive-diagnostic-analytics-understand-big-data/.

17. See "Is It Possible for AI to Be Biased?," *Thomson Reuters Legal Solutions* (blog), March 1, 2018, https://legalsolutions.thomsonreuters.co.uk/blog/2018/03/01/possible-ai-biased/; Paola Cecchi-Dimeglio, "Can We Get the Bias Out of Our AI?," *CPI Antitrust Chronicle*, June 2023, https://www.competitionpolicyinternational.com/wp

-content/uploads/2023/06/4-CAN-WE-GET-THE-BIAS-OUT-OF-OUR-AI-Paola-Cecchi-Dimeglio.pdf.

18. The data presented in the tables reflect the critical insights into biases within AI and large language models (LLMs). Addressing these biases through data diversity, ethical guidelines, and fairness metrics is crucial for building an inclusive and equitable metaverse. The mitigation strategies outlined aim to guide developers, businesses, and policymakers in fostering a more just digital landscape.

19. Donghyun Kim, Subin Oh, and Taeshik Shon, "Digital Forensic Approaches for Metaverse Ecosystems," *Forensic Science International: Digital Investigation* 46, supplement (2023): 301608, https://doi.org/10.1016/j.fsidi.2023.301608; Sofia Marlena Schöbel and Jan Marco Leimeister, "Metaverse Platform Ecosystems," *Electronic Markets* 33, no. 1 (2023): 12, https://doi.org/10.1007/s12525-023-00623-w.

20. J. Clement, "Share of Internet Users Worldwide Who Play Video Games on Any Device as of 2nd Quarter 2024, by Age Group and Gender," Statista, November 11, 2024, https://www.statista.com/statistics/326420/console-gamers-gender/.

21. Courtney Buzzell, Zamir Lalji, Amanda Loyola, Kim Rants, Emily Scofield, and Stephan Zimmermann, "Unlocking Commerce in the Metaverse," McKinsey, June 8, 2023, https://www.mckinsey.com/capabilities/growth-marketing-and-sales/our-insights/unlocking-commerce-in-the-metaverse.

22. The metaverse is indeed already here in some forms, with platforms like Decentraland and The Sandbox providing early examples of what it can offer. However, the vision of the metaverse as a fully integrated, seamless, and all-encompassing digital environment is still in the process of being realized. It's an evolving space, with its full potential yet to be unlocked. So, when we say it must be "ready and working right out of the box," it refers to the expectation that as new features, platforms, and capabilities are introduced, they should be immediately functional and integrate smoothly with existing elements of the metaverse. This is crucial for maintaining user engagement and trust, and for the metaverse to evolve into the all-encompassing digital paradigm it is envisioned to be. Therefore, while elements of the metaverse are currently operational and being used, the broader, more expansive concept of the metaverse—where virtual and physical realities are fully integrated—is still emerging. This dual aspect of the metaverse being both "here" and "still arriving" is a reflection of its ongoing development and the exciting potential it holds for the future.

23. *Framing the Future of Web 3.0* (Goldman Sachs).

24. *Framing the Future of Web 3.0* (Goldman Sachs).

25. Gartner, "Metaverse"; *Metaverse Market Size, Share, Growth Analysis Report, 2030* (Grand View Research).

26. Ivy K. Lau, "Metaverse and Money," *PayPal Newsroom,* June 5, 2023, https://newsroom.paypal-corp.com/2023-06-Metaverse-and-Money; *Beyond Gaming: The*

Real Metaverse Opportunity (KPMG, 2022), https://assets.kpmg.com/content/dam/kpmg/pt/pdf/pt-websummit-kpmg-2022-metaverse-survey-report.pdf.

27. "Metanomics," a term introduced in a course on *Second Life* hosted by Cornell University professor Robert Bloomfield in 2007, aims to provide a fresh approach to business outcomes and analysis.

28. For a more conservative model regarding global metaverse users, see Samarpita Chakraborty, "1.4 Billion Users Expected to Join the Metaverse by 2030," Fintech News, September 26, 2023, https://www.fintechnews.org/1-4-billion-users-expected-to-join-the-metaverse-by-2030/, and for a more liberal predictive model, see Statista's "Metaverse—Worldwide | Statista Market Forecast." Additionally, for the predictive model of US users, refer to "Metaverse—United States | Statista Market Forecast," Statista, updated March 2024, https://www.statista.com/outlook/amo/metaverse/united-states.

29. Kate Sukhanova, "Metaverse Statistics 2024: Latest User & Market Trends," *Techreport*, updated May 27, 2024, https://techreport.com/statistics/Metaverse-statistics/.

30. Gartner, "Metaverse"; *Metaverse Market Size, Share, Growth Analysis Report, 2030* (Grand View Research); "Metaverse—United States | Statista Market Forecast"; "Metaverse—Worldwide | Statista Market Forecast."

31. Metaversed, "The Metaverse Reaches 400m Monthly Active Users," *LinkedIn Blog*, March 12, 2022, https://www.linkedin.com/pulse/Metaverse-reaches-400m-monthly-active-users-Metaversed/?trk=pulse-article_more-articles_related-content-c.

32. J. Clement, "US Adults on Using the Metaverse 2022, by Ethnicity," Statista, February 8, 2023, https://www.statista.com/statistics/1302555/us-adults-using-on-the-Metaverse-by-ethnicity/.

33. J. Clement, "US Adults Who Have Heard of the Metaverse 2022, by Gender," Statista, February 8, 2023, https://www.statista.com/statistics/1302643/us-adults-heard-about-the-Metaverse-gender/.

34. Brandessence Market Research and Consulting Private Limited, "Metaverse Real Estate Market Is Growing at 31.2% CAGR to 2028 Says Brandessence Market Research," *GlobeNewswire by Notified*, January, 25, 2022, https://www.globenewswire.com/news-release/2022/01/25/2372332/0/en/Metaverse-Real-Estate-Market-is-Growing-at-31-2-CAGR-to-2028-Says-Brandessence-Market-Research.html; Debra Kamin, "Investors Snap Up Metaverse Real Estate in a Virtual Land Boom," *New York Times*, November 30, 2021, https://www.nytimes.com/2021/11/30/business/metaverse-real-estate.html.

35. Christine Moy and Adit Gadgil, *Opportunities in the Metaverse* (JPMorgan Chase, 2022), https://www.jpmorgan.com/content/dam/jpm/treasury-services/documents/opportunities-in-the-metaverse.pdf.

36. Annelieke Mooij, *Regulating the Metaverse Economy: How to Prevent Money Laundering and the Financing of Terrorism* (Cham, Switzerland: Springer Nature, 2023); Ivy K. Lau, "Metaverse and Money."

37. *Metaverse Market Size, Share, Growth Analysis Report, 2030* (Grand View Research); Sang-Min Park and Young-Gab Kim, "A Metaverse: Taxonomy, Components, Applications, and Open Challenges," *IEEE Access* 10 (2022): 4209–4251, https://doi.org/10.1109/ACCESS.2021.3140175.

38. Anand Shah and Anu Bahri, "Metanomics: Adaptive Market and Volatility Behaviour in Metaverse," preprint, SSRN, September 1, 2022, https://doi.org/10.2139/ssrn.4206410; Stephen A. Atlas, "Inductive Metanomics: Economic Experiments in Virtual Worlds," *Journal of Virtual Worlds Research* 1, no. 1 (2008).

39. Naomi Nix, "Building a Metaverse That's Not Just for White Guys," *Bloomberg*, November 4, 2021, https://www.bloomberg.com/news/newsletters/2021-11-04/building-a-metaverse-that-s-not-just-for-white-guys.

40. Howard Zhong and Mark Hamilton, "Exploring Gender and Race Biases in the NFT Market," *Finance Research Letters* 53 (2023): https://doi.org/10.1016/j.frl.2023.103651; Sandra Upson, "The 10,000 Faces that Launched an NFT Revolution," *Wired*, November 11, 2021, https://www.wired.com/story/the-10000-faces-that-launched-an-nft-revolution/; Misyrlena Egkolfopoulou and Akayla Gardner, "Even in the Metaverse, Not All Identities Are Created Equal," *Bloomberg*, December 6, 2021, https://www.bloomberg.com/news/features/2021-12-06/cryptopunk-nft-prices-suggest-a-diversity-problem-in-the-metaverse.

41. Egkolfopoulou and Gardner, "Even in the Metaverse, Not All Identities Are Created Equal"; Anushree Dave, "NFT Art Market Boom Is Overwhelmingly Benefiting Male Creators," *Bloomberg*, November 9, 2021, https://www.bloomberg.com/news/articles/2021-11-09/nft-crypto-art-market-boom-biggest-sales-going-to-male-artists-women-lag#xj4y7vzkg.

42. Matthew Ball, *The Metaverse: And How It Will Revolutionize Everything* (New York: Liveright, 2022); QuHarrison Terry and Scott "DJ Skee" Keeney, *The Metaverse Handbook: Innovating for the Internet's Next Tectonic Shift* (Hoboken, NJ: John Wiley & Sons, 2022).

43. Josh Howarth, "75+ Metaverse Statistics (New 2024 Data)," *Exploding Topics*, November 22, 2023, https://explodingtopics.com/blog/metaverse-stats; Moy and Gadgil, *Opportunities in the Metaverse*.

44. Duleesha Kulasooriya, Michelle Khoo, and Michelle Tan, *The Metaverse in Asia: Strategies for Accelerating Economic Impact* (Deloitte Center for the Edge, 2022); Kevin Chan, "Global Tech Hubs, Billions towards GDP and More: How the Metaverse Will Shape Economies," World Economic Forum, June 22, 2023, https://www.weforum.org/agenda/2023/06/what-will-be-the-economic-benefits-of-the-metaverse/.

45. Wai Han Lo and Ka Lun Benjamin Cheng, "Does Virtual Reality Attract Visitors? The Mediating Effect of Presence on Consumer Response in Virtual Reality Tourism Advertising," *Information Technology & Tourism* 22, no. 4 (2020): 537–562, https://doi.org/10.1007/s40558-020-00190-2; Sandra Maria Correia Loureiro, João Guerreiro, and Faizan Ali, "20 Years of Research on Virtual Reality and Augmented Reality in Tourism Context: A Text-Mining Approach," *Tourism Management* 77 (2020): 104028, https://doi.org/10.1016/j.tourman.2019.104028.

46. Decentraland is a virtual world that runs on Ethereum blockchain technology. It's a decentralized, user-owned platform where individuals can buy, sell, and develop virtual land, creating an array of experiences from games to social events. In Decentraland, every piece of content, from the land parcels to the in-world activities, is created and governed by its community. This allows for a high degree of personalization and creativity, as users are not only participants but also creators and decision-makers in this digital realm. The Sandbox, similarly, is a virtual world that empowers users through the blockchain. It leverages user-generated content to create a dynamic and ever-evolving space. In The Sandbox, users can create, own, and monetize their gaming experiences using the platform's utility token, SAND. It's a world that combines elements of decentralized finance with gaming and user-generated content, encouraging a community-driven approach where players have true ownership of their creations. Both Decentraland and The Sandbox represent the shift toward more immersive, interactive, and user-centered online experiences. They are not just games or platforms; they are entire ecosystems where the lines between content creators, players, and entrepreneurs blur, paving the way for innovative forms of collaboration, creation, and commerce in the digital world.

47. *Social Implications of the Metaverse* (World Economic Forum, July 2023).

48. MetaMask is a software cryptocurrency wallet that enables users to interact with the Ethereum blockchain. It allows individuals to manage their Ethereum accounts, send and receive Ethereum-based cryptocurrencies and tokens, and securely connect to decentralized applications (dApps) through a web browser extension or mobile app. Developed by ConsenSys, a blockchain software company focusing on Ethereum-based tools and infrastructure, MetaMask serves as a gateway to the decentralized web, providing users with control over their digital assets and online interactions.

49. Chan, "Global Tech Hubs, Billions towards GDP and More."

50. Cecchi-Dimeglio, "Can We Get the Bias Out of Our AI?"; "Ask Dr Paola: Algorithms and Biases in Recruitment," *Thomson Reuters Legal Solutions* (blog), July 13, 2018, https://blogs.thomsonreuters.com/legal-uk/2018/07/13/ask-dr-paola-algorithms-and-biases-in-recruitment/.

Chapter 3

1. QuHarrison Terry and Scott "DJ Skee" Keeney, *The Metaverse Handbook: Innovating for the Internet's Next Tectonic Shift* (Hoboken, NJ: John Wiley & Sons, 2022); Joe Flower, "How to Build a Metaverse," *New Scientist*, October 14, 1995, https://www.newscientist.com/article/mg14819994-000-how-to-build-a-metaverse/.

2. Rosa María Ricoy-Casas, "The Metaverse as a New Space for Political Communication," in *Communication and Applied Technologies*, ed. Paulo Carlos López-López, Daniel Barredo, Ángel Torres-Toukoumidis, Andrea De-Santis, and Óscar Avilés, vol. 318, Smart Innovation, Systems and Technologies (Singapore: Springer Nature Singapore, 2023), 325–334, https://doi.org/10.1007/978-981-19-6347-6_29.

3. Mary Aiken, *The Cyber Effect: A Pioneering Cyberpsychologist Explains How Human Behavior Changes Online* (New York: Spiegel & Grau, 2016); Grainne Kirwan, "A Battle Cry Echoing from Media Portals: Review of *The Cyber Effect*, by M. Aiken," *The Psychologist* 30 (2017): 67, https://www.bps.org.uk/psychologist/battle-cry-echoing-media-portals.

4. Rhonda Hadi, Shiri Melumad, and Eric S. Park, "The Metaverse: A New Digital Frontier for Consumer Behavior," *Journal of Consumer Psychology* 34, no. 1 (2024): 142–166, https://doi.org/10.1002/jcpy.1356; Eugy Han et al., "People, Places, and Time: A Large-Scale, Longitudinal Study of Transformed Avatars and Environmental Context in Group Interaction in the Metaverse," *Journal of Computer-Mediated Communication* 28, no. 2 (2023): https://doi.org/10.1093/jcmc/zmac031.

5. See tables in the appendix on Types of Social Biases in LLMs, Addressing Social Biases in LLMs, Key Areas to Investigate for Bias in LLMs, Key Questions to Investigate for Biases, and Findings in the Process of Biases in NLP.

6. Maria Kovacova, Veronika Machova, and Daniel Bennett, "Immersive Extended Reality Technologies, Data Visualization Tools, and Customer Behavior Analytics in the Metaverse Commerce," *Journal of Self-Governance and Management Economics* 10, no. 2 (2022): 7–21, https://doi.org/10.22381/jsme10220221.

7. Atiye Pinar Zumrut, "Evaluation of the Metaverse Universe in Light of Psychology and Sociology," in *Metaverse: Technologies, Opportunities, and Threats*, ed. Fatih Sinan Esen, Hasan Tinmaz, and Madhusudan Singh, vol. 133, Studies in Big Data (Singapore: Springer Nature Singapore, 2023), 203–218, https://doi.org/10.1007/978-981-99-4641-9_14.

8. OECD, *How's Life in the Digital Age? Opportunities and Risks of the Digital Transformation for People's Well-Being* (Paris: OECD Publishing, 2019), https://doi.org/10.1787/9789264311800-en.

9. Daniel Kardefelt-Winther, "How Does the Time Children Spend Using Digital Technology Impact Their Mental Well-Being, Social Relationships and Physical

Activity? An Evidence-Focused Literature Review," preprint, Innocenti Discussion Papers, December 31, 2017, https://doi.org/10.18356/cfa6bcb1-en.

10. Amy Edmondson, "Psychological Safety and Learning Behavior in Work Teams," *Administrative Science Quarterly* 44, no. 2 (1999): 350–383, https://doi.org/10.2307/2666999.

11. James Brown, Jeremy Bailenson, and Jeffrey Hancock, "Misinformation in Virtual Reality," *Journal of Online Trust and Safety* 1, no. 5 (2023), https://doi.org/10.54501/jots.v1i5.120; Jinxia Wang, Stanislav Makowski, Alan Cieślik, Haibin Lv, and Zhihan Lv, "Fake News in Virtual Community, Virtual Society, and Metaverse: A Survey," *IEEE Transactions on Computational Social Systems* 11, no. 4 (2024): 4828–4842, https://doi.org/10.1109/TCSS.2022.3220420.

12. Philip Lindner et al., "Experiences of Gamified and Automated Virtual Reality Exposure Therapy for Spider Phobia: Qualitative Study," *JMIR Serious Games* 8, no. 2 (2020): e17807, https://doi.org/10.2196/17807; Theodore Oing and Julie Prescott, "Implementations of Virtual Reality for Anxiety-Related Disorders: Systematic Review," *JMIR Serious Games* 6, no. 4 (2018): e10965, https://doi.org/10.2196/10965; Philip Lindner et al., "Creating State of the Art, Next-Generation Virtual Reality Exposure Therapies for Anxiety Disorders Using Consumer Hardware Platforms: Design Considerations and Future Directions," *Cognitive Behaviour Therapy* 46, no. 5 (2017): 404–420, https://doi.org/10.1080/16506073.2017.1280843.

13. Mike Boland, "VR Usage & Consumer Attitudes (New Report)," VR/AR Association (VRARA), June 24, 2019, https://www.thevrara.com/blog2/2019/2/20/will-ars-killer-app-be-social-new-report-rwccm-g8ajl.

14. Joshua A. Sipper, *The Cyber Meta-Reality: Beyond the Metaverse* (Lanham, MD: Rowman & Littlefield, 2022).

15. Momentum Worldwide, "80% of People Feel More Included in the Metaverse than in Real Life," *PR Newswire*, July 8, 2022, https://www.prnewswire.com/news-releases/80-of-people-feel-more-included-in-the-metaverse-than-in-real-life-301583006.html.

16. Ruoyu Zhao, Yushu Zhang, Youwen Zhu, Rushi Lan, and Zhongyun Hua, "Metaverse: Security and Privacy Concerns," *Journal of Metaverse* 3, no. 2 (2023): 93–99, https://doi.org/10.57019/jmv.1286526; Yogesh K. Dwivedi et al., "Metaverse Beyond the Hype: Multidisciplinary Perspectives on Emerging Challenges, Opportunities, and Agenda for Research, Practice and Policy," *International Journal of Information Management* 66 (2022): 102542, https://doi.org/10.1016/j.ijinfomgt.2022.102542; Roberto Di Pietro and Stefano Cresci, "Metaverse: Security and Privacy Issues," in *2021 Third IEEE International Conference on Trust, Privacy and Security in Intelligent Systems and Applications (TPS-ISA)* (IEEE, 2021), 281–288, https://doi.org/10.1109/TPSISA52974.2021.00032.

17. Suraj Lakhani, "When Digital and Physical World Combine: The Metaverse and Gamification of Violent Extremism," *Perspectives on Terrorism* 17, no. 2 (2023): 108–125, https://doi.org/10.19165/HCZJ7464; Aman Bajwa, "Malevolent Creativity & the Metaverse: How the Immersive Properties of the Metaverse May Facilitate the Spread of a Mass Shooter's Culture," *Journal of Intelligence, Conflict, and Warfare* 5, no. 2 (2022): 32–52, https://doi.org/10.21810/jicw.v5i2.5038; Joel S. Elson, Austin C. Doctor, and Sam Hunter, "The Metaverse Offers a Future Full of Potential—for Terrorists and Extremists, Too," *The Conversation*, January 7, 2022, https://theconversation.com/the-metaverse-offers-a-future-full-of-potential-for-terrorists-and-extremists-too-173622.

18. Tom Boellstorff, "Toward Anthropologies of the Metaverse," *American Ethnologist* 51, no. 1 (2024): 47–56, https://doi.org/10.1111/amet.13228; Moises Perez, Adriana Pineda-Rafols, Maria Pilar Egea-Romero, Maria Gonzalez-Moreno, and Esther Rincon, "Addressing Body Image Disturbance Through Metaverse-Related Technologies: A Systematic Review," *Electronics* 12, no. 22 (2023): 4580, https://doi.org/10.3390/electronics12224580.

19. Carlotta Rigotti and Giancluadio Malgieri, "Human Vulnerability in the Metaverse," *Alliance for Universal Digital Rights*, 2023, https://audri.org/wp-content/uploads/2023/07/EN-AUDRi-Metaverse-Report-07-PDF.pdf; Kuzi Charamba, "Beyond the Corporate Responsibility to Respect Human Rights in the Dawn of a Metaverse," *University of Miami International & Comparative Law Review* 30, no. 1 (2022): 110, https://doi.org/10.2139/ssrn.4043254; Andy Miah, "The Moral Metaverse: Establishing an Ethical Foundation for XR Design," in *Understanding Virtual Reality: Challenging Perspectives for Media Literacy and Education*, ed. Sarah Jones, Steve Dawkins, and Julian McDougall (Oxfordshire, UK: Routledge, 2022), 112–128, https://api.taylorfrancis.com/content/chapters/edit/download?identifierName=doi&identifierValue=10.4324/9780367337032-12&type=chapterpdf.

20. Paola Cecchi-Dimeglio, *Diversity Dividend: The Transformational Power of Small Changes to Debias Your Company, Attract Diverse Talent, Manage Everyone Better—and Make More Money* (Cambridge, MA: MIT Press, 2023); Iris Bohnet, *What Works: Gender Equality by Design*, ill. ed. (Cambridge, MA: Harvard University Press, 2016); Elinor Ostrom, *Understanding Institutional Diversity* (Princeton, NJ: Princeton University Press, 2005), https://doi.org/10.2307/j.ctt7s7wm.

21. Fernanda Herrera and Jeremy N. Bailenson, "Virtual Reality Perspective-Taking at Scale: Effect of Avatar Representation, Choice, and Head Movement on Prosocial Behaviors," *New Media & Society* 23, no. 8 (2021): 2189–2209, https://doi.org/10.1177/1461444821993121; Sachiyo Ueda, Kazuya Nagamachi, Junya Nakamura, Maki Sugimoto, Masahiko Inami, and Michiteru Kitazaki, "The Effects of Body Direction and Posture on Taking the Perspective of a Humanoid Avatar in a Virtual Environment," *PLoS One* 16, no. 12 (2021): e0261063, https://doi.org/10.1371/journal.pone.0261063; Carolyn A. Meyer, Kenneth R. Ivie Jr., Natascha Heise, Katie Brown,

and Tod R. Clapp, "Assessing Understanding of Structural Relationships: Early Lessons in a Large Scale Virtual Reality Deployment," *FASEB* 34, no. S1 (2020): 1, https://doi.org/10.1096/fasebj.2020.34.s1.09804.

22. Hyun-Woo Lee, Kun Chang, Jun-Phil Uhm, and Emmaculate Owiro, "How Avatar Identification Affects Enjoyment in the Metaverse: The Roles of Avatar Customization and Social Engagement," *Cyberpsychology, Behavior, and Social Networking* 26, no. 4 (2023): 255–262, https://doi.org/10.1089/cyber.2022.0257; Mónica Cruz, Abílio Oliveira, and Alessandro Pinheiro, "Meeting Ourselves or Other Sides of Us?—Meta-Analysis of the Metaverse," *Informatics* 10, no. 2 (2023): 47, https://doi.org/10.3390/informatics10020047; Eugy Han, Mark Roman Miller, Nilam Ram, Kristine L. Nowak, and Jeremy N. Bailenson, "Understanding Group Behavior in Virtual Reality: A Large-Scale, Longitudinal Study in the Metaverse," paper presented at the 72nd Annual International Communication Association Conference, Paris, France, May 18, 2022, https://papers.ssrn.com/sol3/papers.cfm?abstract_id=4110154.

23. Portia Wang, Mark R. Miller, Eugy Han, Cyan DeVeaux, and Jeremy N. Bailenson, "Understanding Virtual Design Behaviors: A Large-Scale Analysis of the Design Process in Virtual Reality," *Design Studies* 90 (2024): 101237, https://doi.org/10.1016/j.destud.2023.101237.

24. Miguel Barreda-Ángeles and Tilo Hartmann, "Hooked on the Metaverse? Exploring the Prevalence of Addiction to Virtual Reality Applications," *Frontiers in Virtual Reality* 3 (2022): 1–9, https://doi.org/10.3389/frvir.2022.1031697; Ljubisa Bojic, "Metaverse Through the Prism of Power and Addiction: What Will Happen When the Virtual World Becomes More Attractive than Reality?," *European Journal of Futures Research* 10, no. 1 (2022): 1–24, https://doi.org/10.1186/s40309-022-00208-4.

Chapter 4

1. Ben Cormier, "Analyzing If and How International Organizations Contribute to the Sustainable Development Goals: Combining Power and Behavior," *Journal of Organizational Behavior* 39, no. 5 (2018): 545–558, https://doi.org/10.1002/job.2163.

2. Carlo Borzaga and Riccardo Bodini, "What to Make of Social Innovation? Towards a Framework for Policy Development," *Social Policy and Society* 13, no. 3 (2014): 411–421, http://doi.org/10.1017/S1474746414000116.

3. QuHarrison Terry and Scott "DJ Skee" Keeney, *The Metaverse Handbook: Innovating for the Internet's Next Tectonic Shift* (Hoboken, NJ: John Wiley & Sons, 2022); Tambiama Madiega, Polona Car, and Maria Niestadt, *Metaverse: Opportunities, Risks and Policy Implications* (European Parliament, Members' Research Service, 2022), https://www.europarl.europa.eu/RegData/etudes/BRIE/2022/733557/EPRS_BRI(2022)733557_EN.pdf; *Rolling Plan for ICT Standardisation* (European Commission, 2023), https://interoperable-europe.ec.europa.eu/collection/rolling-plan-ict-standardisation

/metaverse; "ITU Secretary-General Outlines an Inclusive Digital Future," World Economic Forum, January 22, 2024, https://www.itu.int/hub/2024/01/wef-2024-itu-secretary-general-outlines-an-inclusive-digital-future/; Hikmet Ersek, "A Reality Check on Inclusive Innovation," World Economic Forum, January 13, 2020, https://www.weforum.org/stories/2020/01/reality-check-inclusive-innovation/; European Commission, *Communication from the Commission to the European Parliament, the Council, the European Economic and Social Committee and the Committee of the Regions: An EU Initiative on Web 4.0 and Virtual Worlds: A Head Start in the Next Technological Transition*, COM(2023) 442 final, July 11, 2023, https://eur-lex.europa.eu/legal-content/EN/TXT/?uri=celex:52023DC0442.

4. Natalie Marchant, "8 Charts that Show the Impact of Race and Gender on Technology Careers," World Economic Forum, April 13, 2021, https://www.weforum.org/agenda/2021/04/gender-race-tech-industry/; Matt Gonzales, "Racial, Gender Discrimination in Tech Industry Worsening," *Society for Human Resource Management*, https://www.shrm.org/resourcesandtools/hr-topics/behavioral-competencies/global-and-cultural-effectiveness/pages/racial-gender-discrimination-in-tech-industry-worsening.aspx; *Special Report: Diversity in Tech* (Washington, DC: US Equal Employment Opportunity Commission, 2016), https://www.eeoc.gov/special-report/diversity-high-tech; *High Tech, Low Inclusion: Diversity in the High Tech Workforce and Sector, 2014–2022* (Washington, DC: US Equal Employment Opportunity Commission, 2024), https://www.eeoc.gov/sites/default/files/2024-09/20240910_Diversity%20in%20the%20High%20Tech%20Workforce%20and%20Sector%202014-2022.pdf; Cindy Brown Barnes, *Diversity in the Technology Sector: Federal Agencies Could Improve Oversight of Equal Employment Opportunity Requirements* (Washington, DC: US Government Accountability Office, 2017), https://www.gao.gov/assets/gao-18-69.pdf.

5. "Civilian Labor Force, by Age, Sex, Race, and Ethnicity," US Bureau of Labor Statistics (BLS), last modified August 29, 2024, https://www.bls.gov/emp/tables/civilian-labor-force-summary.htm; "Labor Force Statistics from the Current Population Survey," US BLS, last modified December 31, 2024, https://www.bls.gov/cps/demographics.htm; *Special Report* (US Equal Employment Opportunity Commission); *High Tech, Low Inclusion* (US Equal Employment Opportunity Commission).

6. *Special Report* (US Equal Employment Opportunity Commission); *High Tech, Low Inclusion* (US Equal Employment Opportunity Commission).

7. *Special Report* (US Equal Employment Opportunity Commission); *High Tech, Low Inclusion* (US Equal Employment Opportunity Commission).

8. *U.S. Equal Employment Opportunity Commission (EEOC) Employer Information Report (EEO-1 Component 1)* (EEOC, 2023), https://query.prod.cms.rt.microsoft.com/cms/api/am/binary/RW1hym2.

9. *Embracing Change Through Inclusion: Meta's 2022 Diversity Report* (Menlo Park, CA: Meta, 2022), https://about.fb.com/wp-content/uploads/2022/07/Meta_Embracing

-Change-Through-Inclusion_2022-Diversity-Report.pdf; *Equal Employment Opportunity 2021 Employer Information Report EEO-1* (Menlo Park, CA: Meta, 2022), https://about.fb.com/wp-content/uploads/2022/07/2021-Equal-Employment-Opportunity-Report.pdf; *Equal Employment Opportunity 2020 Employer Information Report EEO-1* (Menlo Park, CA: Meta, 2021), https://about.fb.com/wp-content/uploads/2021/07/EEO-1_Report.pdf.

The EEO-1 Component 1 report is a mandatory annual data collection for private-sector employers with 100 or more employees and federal contractors with 50 or more employees who meet specific criteria. Employers must submit workforce demographic data, including job category, sex, and race or ethnicity, to the EEOC. This data collection is authorized under Section 709(c) of Title VII of the Civil Rights Act of 1964, as amended (42 U.S.C. 2000e, et seq.), Sections 1602.7–1602.14, Chapter XIV, Title 29 of the *Code of Federal Regulations* (CFR), Exec. Order No. 11246, 30 Fed. Reg. 12319 (Sept. 24, 1965), and 41 C.F.R. 60-1.7(a).

For historical reports and archived resource materials, companies can access the EEO-1 component 1 website at www.eeocdata.org/eeo1. Updates regarding the 2024 EEO-1 component 1 data collection, including the opening date, will also be published on this site.

10. *Google Belonging: 2022 Diversity Annual Report* (Menlo Park, CA: Google, 2022), https://about.google/belonging/diversity-annual-report/2022/; "Inclusion and Diversity," Apple, December 2022, https://www.apple.com/diversity/.

11. Giovanni Russonello, "How Big Tech Allows the Racial Wealth Gap to Persist," *New York Times*, June 21, 2021, https://www.nytimes.com/2021/06/21/us/politics/big-tech-racial-wealth-gap.html.

12. Maria Elena Baltazar Herrera, "Innovation for Impact: Business Innovation for Inclusive Growth," *Journal of Business Research* 69, no. 5 (2016): 1725–1730, https://doi.org/10.1016/j.jbusres.2015.10.045; Yogesh K. Dwivedi et al., "Metaverse Beyond the Hype: Multidisciplinary Perspectives on Emerging Challenges, Opportunities, and Agenda for Research, Practice and Policy," *International Journal of Information Management* 66 (2022): 102542, https://doi.org/10.1016/j.ijinfomgt.2022.102542; Greg Schrock and Nichola Lowe, "Inclusive Innovation Editorial: The Promise of Inclusive Innovation," *Local Economy* 36, no. 3 (2021): 181–186, https://doi.org/10.1177/02690942211042254.

13. Tables 3 and 4 in the appendix provide a comprehensive breakdown of how these biases manifest in AI systems and suggest strategies for mitigating their impact in virtual settings.

14. Daniel Kahneman, *Thinking, Fast and Slow* (New York: Farrar, Straus and Giroux, 2011); Daniel Kahneman, "A Perspective on Judgment and Choice: Mapping Bounded Rationality," *American Psychologist* 58, no. 9 (2003): 697–720, https://doi.org/10.1037/0003-066X.58.9.697; Daniel Kahneman, "Reference Points, Anchors,

Norms, and Mixed Feelings," *Organizational Behavior and Human Decision Processes* 51, no. 2 (1992): 296–312, https://doi.org/10.1016/0749-5978(92)90015-Y.

15. Giorgio A. Tasca, "Twenty-Five Years of *Group Dynamics: Theory, Research and Practice*: Introduction to the Special Issue," *Group Dynamics: Theory, Research, and Practice* 25, no. 3 (2021): 205–212, https://doi.org/10.1037/gdn0000167.

16. Mahzarin R. Banaji and Anthony G. Greenwald, *Blindspot: Hidden Biases of Good People*, rep. ed. (New York: Bantam, 2016).

17. Paola Cecchi-Dimeglio, "How Leaders Are Making Decisions and Connecting the Dots," *Forbes*, July 25, 2023, https://www.forbes.com/sites/paolacecchi-dimeglio/2023/07/25/how-leaders-are-making-decisions-and-connecting-the-dots/.

18. Paola Cecchi-Dimeglio, "Avoiding Flawed Decisions: How Leaders Can Overcome Cognitive Biases," *Forbes*, July 29, 2023, https://www.forbes.com/sites/paolacecchi-dimeglio/2023/07/29/avoiding-flawed-decisions-how-leaders-can-overcome-cognitive-biases/.

19. Joyce Ehrlinger, Ainsley L. Mitchum, and Carol S. Dweck, "Understanding Overconfidence: Theories of Intelligence, Preferential Attention, and Distorted Self-Assessment," *Journal of Experimental Social Psychology* 63 (2016): 94–100, https://doi.org/10.1016/j.jesp.2015.11.001.

20. Howard Garland and Stephanie Newport, "Effects of Absolute and Relative Sunk Costs on the Decision to Persist with a Course of Action," *Organizational Behavior and Human Decision Processes* 48, no. 1 (1991): 55–69, https://doi.org/10.1016/0749-5978(91)90005-E; Daniel Kahneman and Amos Tversky, "Prospect Theory: An Analysis of Decision under Risk," *Econometrica* 47, no. 2 (1979): 263, https://doi.org/10.2307/1914185.

21. Kathryn Whitenton, "Decision Frames: How Cognitive Biases Affect UX Practitioners," Nielsen Norman Group, updated July 16, 2024, https://www.nngroup.com/articles/decision-framing-cognitive-bias-ux-pros/; Katherine L. Milkman, Modupe Akinola, and Dolly Chugh, "What Happens Before? A Field Experiment Exploring How Pay and Representation Differentially Shape Bias on the Pathway into Organizations," *Journal of Applied Psychology* 100, no. 6 (2015): 1678–1712, https://doi.org/10.1037/apl0000022; Jack B. Soll, Katherine L. Milkman, and John W. Payne, "Outsmart Your Own Biases," *Harvard Business Review* 93, no. 5 (2015): 64–71.

22. James Felton, "Why Are Women More Likely to Die in Car Crashes Than Men?," *IFLScience*, February 1, 2023, https://www.iflscience.com/why-are-women-more-likely-to-die-in-car-crashes-than-men-67353.

23. Rüdiger F. Pohl, ed., *Cognitive Illusions: A Handbook on Fallacies and Biases in Thinking, Judgement and Memory* (London: Psychology Press, 2012).

24. Richard E. Nisbett and Timothy D. Wilson, "The Halo Effect: Evidence for Unconscious Alteration of Judgments," *Journal of Personality and Social Psychology* 35, no. 4 (1977): 250–256, https://doi.org/10.1037/0022-3514.35.4.250; E. L. Thorndike, "A Constant Error in Psychological Ratings," *Journal of Applied Psychology* 4, no. 1 (1920): 25–29, https://doi.org/10.1037/h0071663.

25. Daniel R. Stalder, *The Power of Context: How to Manage Our Bias and Improve Our Understanding of Others* (Amherst, NY: Prometheus Books, 2018); Jennifer Inauen, Patrick E. Shrout, Niall Bolger, Gertraud Stadler, and Urte Scholz, "Mind the Gap? An Intensive Longitudinal Study of Between-Person and Within-Person Intention–Behavior Relations," *Annals of Behavioral Medicine* 50, no. 4 (2016): 516–522, https://doi.org/10.1007/s12160-016-9776-x.

26. Juyeon Park, "The Effect of Virtual Avatar Experience on Body Image Discrepancy, Body Satisfaction and Weight Regulation Intention," *Cyberpsychology* 12, no. 1 (2018): 20, https://doi.org/10.5817/CP2018-1-3; Juyeon Park, "Emotional Reactions to the 3D Virtual Body and Future Willingness: The Effects of Self-Esteem and Social Physique Anxiety," *Virtual Reality* 22, no. 1 (2018): 1–11, https://doi.org/10.1007/s10055-017-0314-3.

27. Juyeon Park and Jennifer Paff Ogle, "How Virtual Avatar Experience Interplays with Self-Concepts: The Use of Anthropometric 3D Body Models in the Visual Stimulation Process," *Fashion and Textiles* 8 (2021): 1–24, https://doi.org/10.1186/s40691-021-00257-6.

28. Grace Y. S. Leung, Adrian K. T. Ng, and Henry Y. K. Lau, "Effect of Height Perception on State Self-Esteem and Cognitive Performance in Virtual Reality," in *Engineering Psychology and Cognitive Ergonomics: 18th International Conference, EPCE 2021, Held as Part of the 23rd HCI International Conference, HCII 2021, Virtual Event, July 24–29, 2021, Proceedings*, ed. Don Harris and Wen-Chin Li, Lecture Notes in Computer Science (Cham, Switzerland: Springer International, 2021), 12767:172–184, https://doi.org/10.1007/978-3-030-77932-0_15.

29. Margaret A. Nosek, Susan Robinson-Whelen, Rosemary B. Hughes, and Thomas M. Nosek, "An Internet-Based Virtual Reality Intervention for Enhancing Self-Esteem in Women with Disabilities: Results of a Feasibility Study," *Rehabilitation Psychology* 61, no. 4 (2016): 358–370, https://doi.org/10.1037/rep0000107.

30. Vyjayanti T. Desai, Anna Metz, and Jing Lu, "The Global Identification Challenge: Who Are the 1 Billion People without Proof of Identity?," *World Bank: Voices* (blog), April 25, 2018, https://blogs.worldbank.org/voices/global-identification-challenge-who-are-1-billion-people-without-proof-identity.

31. Julia Clark, Anna Metz, and Claire Casher, "850 Million People Globally Don't Have ID—Why This Matters and What We Can Do About It," *World Bank: Digital Transformation* (blog), February 6, 2023, https://blogs.worldbank.org/digital-develop

ment/850-million-people-globally-dont-have-id-why-matters-and-what-we-can-do-about.

32. Douglas Broom, "A Billion People Have No Legal Identity—But a New App Plans to Change That," World Economic Forum, November 20, 2020, https://www.weforum.org/agenda/2020/11/legal-identity-id-app-aid-tech/. For further insights on metaverse identity and the governance track of the Defining and Building the Metaverse initiative, see *Metaverse Identity: Defining the Self in a Blended Reality* (World Economic Forum, March 2024), https://www3.weforum.org/docs/WEF_Metaverse_Identity_Defining_the_Self_in_a_Blended_Reality_2024.pdf.

33. Josh Howarth, "75+ Metaverse Statistics (New 2024 Data)," *Exploding Topics*, November 22, 2023, https://explodingtopics.com/blog/metaverse-stats.

34. See, for instance, the application of the Safe Zone widget for designers in Meta Horizon, "Safe Areas," accessed February 10, 2025, https://developers.meta.com/horizon/resources/publish-safe-areas/; and in Unreal Engine's "UMG Safe Zones" documentation, accessed February 10, 2025, https://dev.epicgames.com/documentation/en-us/unreal-engine/umg-safe-zones-in-unreal-engine.

35. Sheera Frenkel and Kellen Browning, "The Metaverse's Dark Side: Here Come Harassment and Assaults," *New York Times*, December 30, 2021, https://www.nytimes.com/2021/12/30/technology/metaverse-harassment-assaults.html.

36. Paola Cecchi-Dimeglio, Taha Masood, and Andy Ouderkirk, "What Makes Innovation Partnerships Succeed," *Harvard Business Review*, July 14, 2022, https://hbr.org/2022/07/what-makes-innovation-partnerships-succeed.

37. Sara Reardon, "Gender Gap in US Patents Leads to Few Inventions That Help Women," *Nature* 597, no. 7874 (2021): 139–140, https://doi.org/10.1038/d41586-021-02298-9; Dianna G. El Hioum and Gregory Logan, "USPTO Has Ways to Improve Patent Diversity and Inclusion," *Bloomberg Law*, October 13, 2021, https://news.bloomberglaw.com/ip-law/uspto-has-ways-to-improve-patent-diversity-and-inclusion.

Chapter 5

1. Patrick Grother, Mei Ngan, and Kayee Hanaoka, *Face Recognition Vendor Test (FVRT)—Part 3: Demographic Effects* (Gaithersburg, MD: National Institute of Standards and Technology, 2019), https://doi.org/10.6028/NIST.IR.8280; Joy Buolamwini and Timnit Gebru, "Gender Shades: Intersectional Accuracy Disparities in Commercial Gender Classification," *Proceedings of Machine Learning Research* 81 (2018): 77–91, https://proceedings.mlr.press/v81/buolamwini18a.html.

2. *U.S. Self-Checkout Systems Market | Industry Report, 2030* (Grand View Research, 2022), https://www.grandviewresearch.com/industry-analysis/us-self-checkout-systems-market-report; The National Federation of the Blind, Inc. et al v. Wal-Mart Associates,

Inc., No. 1:2018cv03301—Document 124 (D. Md. 2021), https://law.justia.com/cases/federal/district-courts/maryland/mddce/1:2018cv03301/435932/124/; Lighthouse for the Blind & Visually Impaired v. Redbox Automated Retail, LLC, No. C 12–0195 PJH (N.D. Cal. May 18, 2012), https://casetext.com/case/lighthouse-for-the-blind-visually-impaired-v-redbox-automated-retail.

3. Claudia Lopez Lloreda, "How Speech-Recognition Software Discriminates against Minority Voices," *Scientific American*, October 1, 2020, https://www.scientificamerican.com/article/how-speech-recognition-software-discriminates-against-minority-voices/.

4. Artificial intelligence (AI), large language models (LLMs), and natural language processing (NLP) are powerful tools that offer unprecedented opportunities for inclusivity in the metaverse. These technologies enable more intuitive interactions, break language barriers, and make virtual spaces more accessible by tailoring experiences to diverse user needs. However, they also pose risks of amplifying biases if not designed inclusively. In the appendix, the reader can find a table summary of biases embedded in AI systems that can limit the equitable experience of users and reinforce existing inequalities, as well as strategies to overcome them. The tables do not provide an exhaustive list, but rather, they offer guidance on getting on the right track.

5. Piyush Tantia, "Behavioral Science for Impact: The New Science of Designing for Humans," in *Perspectives on Impact: Leading Voices on Making Systemic Change in the Twenty-First Century*, ed. Nina Montgomery (London: Routledge, 2019), 100–110, https://doi.org/10.4324/9780429452796-9.

6. Paola Cecchi-Dimeglio, "Diversity Nudges," *MIT Sloan Management Review*, November 21, 2023, https://sloanreview.mit.edu/article/diversity-nudges/.

7. Richard H. Thaler and Cass R. Sunstein, *Nudge: Improving Decisions About Health, Wealth, and Happiness* (New York: Penguin Books, 2009).

8. Alistair Rennie and Jonny Protheroe, *How People Decide What to Buy Lies in the "Messy Middle" of the Purchase Journey* (Think with Google: Consumer Insights, July 2020), https://www.thinkwithgoogle.com/consumer-insights/consumer-journey/navigating-purchase-behavior-and-decision-making/.

9. See Amazon "Research Area: Economics," accessed February 10, 2025, https://www.amazon.science/research-areas/economics; Marcel Das, Peter Ester, and Lars Kaczmirek, eds., *Social and Behavioral Research and the Internet: Advances in Applied Methods and Research Strategies*, European Association of Methodology Series (New York: Routledge, 2011); Aaron J. Moss et al., "Using Market-Research Panels for Behavioral Science: An Overview and Tutorial," *Advances in Methods and Practices in Psychological Science* 6, no. 2 (2023), https://doi.org/10.1177/25152459221140388.

10. Nancy Harhut, *Using Behavioral Science in Marketing: Drive Customer Action and Loyalty by Prompting Instinctive Responses* (New York: Kogan Page, 2022).

11. Ipsos, *Understanding Society: How Do We Change Behaviour? Make It Simple* (London: Ipsos MORI, Social Research Institute, 2013), https://www.ipsos.com/sites/default/files/publication/1970-01/sri-understanding-society-april-2013.pdf.

12. Jeanne Liedtka, "Why Design Thinking Works," *Harvard Business Review* 96, no. 9 (2018), https://hbr.org/2018/09/why-design-thinking-works.

13. Paola Cecchi-Dimeglio, "Uncover the Superpower Guiding Leaders to Success," *Forbes*, December 13, 2023, https://www.forbes.com/sites/paolacecchi-dimeglio/2023/12/13/uncover-the-superpower-guiding-leaders-to-success/?sh=5f8426937181.

14. Laura Furstenthal, Alex Morris, and Erik Roth, "Overcoming Barriers to Innovation," McKinsey, June 3, 2022, https://www.mckinsey.com/capabilities/strategy-and-corporate-finance/our-insights/fear-factor-overcoming-human-barriers-to-innovation.

15. "The Global Gender Gap in Innovation and Creativity," World Intellectual Property Organization, 2024, https://www.wipo.int/about-ip/en/ip_innovation_economics/gender_innovation_gap/gender-parity-patenting.html; Michelle Saksena, Nicholas Rada, and Lisa Cook, *Where Are U.S. Women Patentees? Assessing Three Decades of Growth* (Alexandria, VA: US Patent and Trademark Office, 2022), https://www.uspto.gov/sites/default/files/documents/oce-women-patentees-report.pdf; Sara Reardon, "Gender Gap in US Patents Leads to Few Inventions That Help Women," *Nature* 597, no. 7874 (2021): 139–140, https://doi.org/10.1038/d41586-021-02298-9; Paola Cecchi-Dimeglio, Taha Masood, and Andy Ouderkirk, "What Makes Innovation Partnerships Succeed," *Harvard Business Review*, July 14, 2022, https://hbr.org/2022/07/what-makes-innovation-partnerships-succeed.

16. Dianna G. El Hioum and Gregory Logan, "USPTO Has Ways to Improve Patent Diversity and Inclusion," *Bloomberg Law*, October 13, 2021, https://news.bloomberglaw.com/ip-law/uspto-has-ways-to-improve-patent-diversity-and-inclusion.

17. "The Diversity Pledge," Increasing Diversity in Innovation, accessed February 12, 2025, https://increasingdii.org/pledge/. The author has been involved with several major companies that have signed the pledge to help them implement interventions based on science to increase and diversify the number of patent submissions and holders at their companies.

18. Jane Lu, "How to Address the Diversity Challenges of the Metaverse," World Economic Forum, June 14, 2022, https://www.weforum.org/agenda/2022/06/metaverse-platforms-face-diversity-equity-and-inclusion-challenges-heres-how-to-address-them/.

19. Paola Cecchi-Dimeglio, *Diversity Dividend: The Transformational Power of Small Changes to Debias Your Company, Attract Diverse Talent, Manage Everyone Better—and Make More Money* (Cambridge, MA: MIT Press, 2023); Mustafa Özbilgin and Ahu Tatli, "Mapping Out the Field of Equality and Diversity: Rise of Individualism and

Voluntarism," *Human Relations* 64, no. 9 (2011): 1229–1253, https://doi.org/10.1177/0018726711413620.

20. Tomoko Yokoi, Nikolaus Obwegeser, and Michela Beretta, "How Digital Inclusion Can Help Solve Grand Challenges," *MIT Sloan Management Review*, June 14, 2021, https://sloanreview.mit.edu/article/how-digital-inclusion-can-help-solve-grand-challenges/.

21. "History in Your Hands," *Retro Gamer* no. 63 (2009): 24, https://issuu.com/roylazarovich/docs/retro_gamer_063.

22. Aaron Mok, "Nvidia CEO Says He Wakes Up Every Morning Worried His Company Might Fail," *Business Insider*, November 29, 2023, https://www.businessinsider.com/nvidia-ceo-jensen-huang-always-worrying-about-failure-2023-11.

23. Justin Munafo, Meg Diedrick, and Thomas A. Stoffregen, "The Virtual Reality Head-Mounted Display Oculus Rift Induces Motion Sickness and Is Sexist in Its Effects," *Experimental Brain Research* 235, no. 3 (2017): 889–901, https://doi.org/10.1007/s00221-016-4846-7; Bradley Austin Davis, Karen Bryla, and Phillips Alexander Benton, *Oculus Rift in Action* (Shelter Island, NY: Manning, 2015); Mirjam Vosmeer and Ben Schouten, "Creating Video Content for Oculus Rift: Scriptwriting for 360° Interactive Video Productions," in *Entertainment Computing—ICEC 2015*, Lecture Notes in Computer Science (Cham, Switzerland: Springer International, 2015), 9353:556–559, https://doi.org/10.1007/978-3-319-24589-8_56.

24. Kati Alha, Elina Koskinen, Janne Paavilainen, and Juho Hamari, "Why Do People Play Location-Based Augmented Reality Games: A Study on Pokémon GO," *Computers in Human Behavior* 93 (2019): 114–122, https://doi.org/10.1016/j.chb.2018.12.008; Katherine B. Howe, Christian Suharlim, Peter Ueda, Daniel Howe, Ichiro Kawachi, and Eric B. Rimm, "Gotta Catch'em All! Pokémon GO and Physical Activity among Young Adults: Difference in Differences Study," *BMJ* 355 (2016): i6270, https://doi.org/10.1136/bmj.i6270.

25. Quinn Myers, *Google Glass*, Remember the Internet, vol. 3 (New York: Instar Books, 2022); Amarolinda Klein, Carsten Sørensen, Angilberto Sabino de Freitas, Cristiane Drebes Pedron, and Silvia Elaluf-Calderwood, "Understanding Controversies in Digital Platform Innovation Processes: The Google Glass Case," *Technological Forecasting & Social Change* 152 (2020): 119883, https://doi.org/10.1016/j.techfore.2019.119883; Allen Firstenberg and Jason Salas, *Designing and Developing for Google Glass: Thinking Differently for a New Platform* (Sebastopol, CA: O'Reilly Media, 2015); John A. M. Paro, Rahim Nazareli, Anadev Gurjala, Aaron Berger, and Gordon K. Lee, "Video-Based Self-Review: Comparing Google Glass and GoPro Technologies," *Annals of Plastic Surgery* 74, supplement 1 (2015): S71–74, https://doi.org/10.1097/SAP.0000000000000423; Katherine J. Klein and Joann Speer Sorra, "The Challenge of Innovation Implementation," *Academy of Management Review* 21, no. 4 (1996): 1055–1080, https://doi.org/10.2307/259164.

26. Amy Edmondson, *Right Kind of Wrong: The Science of Failing Well* (New York: Simon Element, 2023).

27. Alessandro Narduzzo and Valentina Forrer, "Nurturing Innovation Through Intelligent Failure: The Art of Failing on Purpose," *Technovation* 131 (2024): 102951, https://doi.org/10.1016/j.technovation.2024.102951; Mark D. Cannon and Amy C. Edmondson, "Failing to Learn and Learning to Fail (Intelligently): How Great Organizations Put Failure to Work to Innovate and Improve," *Long Range Planning* 38, no. 3 (2005): 299–319, https://doi.org/10.1016/j.lrp.2005.04.005.

28. Carol S. Dweck, *Mindset: The New Psychology of Success*, Ballantine Books trade paperback ed. (New York: Ballantine Books, 2007); Carol S. Dweck, *Essays in Social Psychology—Self-Theories: Their Role in Motivation, Personality, and Development* (Oxfordshire, UK: Psychology Press, 1999), https://doi.org/10.4324/9781315783048.

29. Artur Dias and Aurora A. C. Teixeira, "The Anatomy of Business Failure: A Qualitative Account of Its Implications for Future Business Success," *European Journal of Management and Business Economics* 26, no. 1 (2017): 2–20, https://doi.org/10.1108/EJMBE-07-2017-001; Gerard P. Hodgkinson and George Wright, "Confronting Strategic Inertia in a Top Management Team: Learning from Failure," *Organization Studies* 23, no. 6 (2002): 949–977, https://doi.org/10.1177/0170840602236014.

30. Cary Lu, "The PDA Comeback? (Evaluations of Four Personal Digital Assistants) (Includes Related Articles About Working on PDAs and the Macintosh Newton's Handwriting Recognition Capability) (Hardware Review)," *Macworld* (San Francisco: IDG Consumer & SMB, Inc, 1996); Dave Andrews, "Personal Digital Assistants: Behind the Wheel of the First Zoomer and Newton PDAs," *Byte* 18, no. 10 (1993): 22.

31. Brad Kelechava, "Blu-ray vs HD DVD: Standard Format War," *American National Standards Institute* (blog), May 13, 2016, https://blog.ansi.org/blu-ray-vs-hd-dvd-standard-format-war/.

32. Jason Ward, "Former and Current Microsoft Staffers Talk About Why Windows Phones Failed," Windows Central, last updated August 12, 2020, https://www.windowscentral.com/microsofts-terry-myerson-and-others-why-windows-phone-failed-thats-fixed-now.

33. Zilin Wang and Moon-Tong Chan, "A Systematic Review of Google Cardboard Used in Education," *Computers & Education: X Reality* 4 (2024): 100046, https://doi.org/10.1016/j.cexr.2023.100046; Kang Hao Cheong, Joel Weijia Lai, Jun Hong Yap, Gideon Sian Wee Cheong, Stephanie Vericca Budiman, and Omar Ortiz, "Utilising Google Cardboard Virtual Reality for Visualization in Multivariable Calculus," *IEEE Access* 11 (2023): 75398–75406, https://doi.org/10.1109/ACCESS.2023.3281753; Wendy Powell, Vaughan Powell, Phillip Brown, Marc Cook, and Jahangir Uddin, "Getting Around in Google Cardboard—Exploring Navigation Preferences with Low-Cost Mobile VR," in *2016 IEEE 2nd Workshop on Everyday Virtual Reality (WEVR)* (IEEE, 2016), 5–8, https://doi.org/10.1109/WEVR.2016.7859536.

34. Sarah Fields, "Oculus Is Discontinuing the Oculus Go Standalone VR Headset," *GameRant*, June 24, 2020, https://gamerant.com/oculus-go-virtual-reality-headset-discontinued/.

35. Ian Sherr, "Say Goodbye to Google's Daydream VR Experiment," *CNET*, October 15, 2019, https://www.cnet.com/tech/mobile/say-goodbye-to-googles-daydream-vr-experiment/.

36. John Elmore, "Why Is Vive Discontinued: Exploring the Rise and Fall of HTC's VR Headset," *TheTechyLife*, April 15, 2024, https://thetechylife.com/why-is-vive-discontinued/.

37. "History | Company Information," I-O DATA, accessed March 6, 2024, https://www.iodata.com/company/history.htm.

38. Chris Wiltz, "The Story of Sega VR: Sega's Failed Virtual Reality Headset," *Design News*, March 1, 2019, https://www.designnews.com/testing-measurement/the-story-of-sega-vr-sega-s-failed-virtual-reality-headset.

39. Adi Robertson, "Augmented Reality Headset Company Daqri Is Reportedly Shutting Down," *The Verge*, September 13, 2019, https://www.theverge.com/2019/9/13/20864556/daqri-ar-headset-smart-glasses-startup-shutdown-asset-sale-layoffs; Reena Mukamal, "Are Virtual Reality Headsets Safe for Eyes?," American Academy of Ophthalmology (AAO), August 21, 2024, https://www.aao.org/eye-health/tips-prevention/are-virtual-reality-headsets-safe-eyes; Julie Iskander, Mohammed Hossny, and Saeid Nahavandi, "Using Biomechanics to Investigate the Effect of VR on Eye Vergence System," *Applied Ergonomics* 81 (2019), https://www.sciencedirect.com/science/article/abs/pii/S0003687018302904.

40. Daniel Kahneman, *Thinking, Fast and Slow* (New York: Farrar, Straus and Giroux, 2011); Daniel Kahneman and Amos Tversky, "Prospect Theory: An Analysis of Decision under Risk," *Econometrica* 47, no. 2 (1979): 263, https://doi.org/10.2307/1914185.

41. Amy Edmondson, "Psychological Safety and Learning Behavior in Work Teams," *Administrative Science Quarterly* 44, no. 2 (1999): 350–383, https://doi.org/10.2307/2666999.

42. Titus Winters, Tom Manshreck, and Hyrum Wright, *Software Engineering at Google: Lessons Learned from Programming Over Time* (Beijing: O'Reilly Media, 2020); Charles Duhigg, *Smarter Faster Better: The Transformative Power of Real Productivity* (New York: Random House, 2016).

43. Teresa M. Amabile, Colin M. Fisher, and Julianna Pillemer, "IDEO's Culture of Helping," *Harvard Business Review* (2014), https://hbr.org/2014/01/ideos-culture-of-helping.

44. Amy C. Edmondson, *The Fearless Organization: Creating Psychological Safety in the Workplace for Learning, Innovation, and Growth* (Hoboken, NJ: John Wiley & Sons, 2018).

45. Paola Cecchi-Dimeglio, "Men as Gender Allies: Verizon's Craig Silliman and Walmart's Alan Bryan," in *Global Champions of Sustainable Development*, ed. Patricia M. Flynn, Milenko Gudić, and Tay Keong Tan (London: Routledge, 2019), https://doi.org/10.4324/9781351176316-13.

46. Paola Cecchi-Dimeglio, "8 Characteristics of an Effective Leader," *Forbes*, November 28, 2023, https://www.forbes.com/sites/paolacecchi-dimeglio/2023/11/28/8-characteristics-of-an-effective-leader/.

47. Eugene Kim, "How Amazon CEO Jeff Bezos Thinks About Failure," *Business Insider*, May 28, 2016, https://www.businessinsider.com/how-amazon-ceo-jeff-bezos-thinks-about-failure-2016-5?international=true&r=US&IR=T; Taylor Soper, "'Failure and Innovation Are Inseparable Twins': Amazon Founder Jeff Bezos Offers 7 Leadership Principles," *GeekWire*, October 28, 2016, https://www.geekwire.com/2016/amazon-founder-jeff-bezos-offers-6-leadership-principles-change-mind-lot-embrace-failure-ditch-powerpoints/.

48. Marcel Schwantes, "Elon Musk: Tesla and SpaceX Had Only 10 Percent Chance of Success," *Inc.*, January 15, 2020, https://www.inc.com/erik-sherman/elon-musk-tesla-spacex-success.html; Matt Pressman, "Elon Musk: You Should Be Failing. 'If Things Are Not Failing, You Are Not Innovating Enough,'" *CleanTechnica*, February 1, 2020, https://cleantechnica.com/2020/02/01/elon-musk-you-should-be-failing-if-things-are-not-failing-you-are-not-innovating-enough/.

49. Cecchi-Dimeglio, "8 Characteristics of an Effective Leader"; Carmine Gallo, "How Great Leaders Communicate," *Harvard Business Review*, November 23, 2022, https://hbr.org/2022/11/how-great-leaders-communicate.

50. Paola Cecchi-Dimeglio, "In Times of Anxiety, Lead With 'We' and 'Us,'" *MIT Sloan Management Review*, May 18, 2020, https://sloanreview.mit.edu/article/in-times-of-anxiety-lead-with-we-and-us/.

51. Paola Cecchi-Dimeglio, "Why Sharing Good News Matters," *MIT Sloan Management Review*, June 17, 2020, https://sloanreview.mit.edu/article/why-sharing-good-news-matters/.

52. Alex Bell, Raj Chetty, Xavier Jaravel, Neviana Petkova, and John Van Reenen, "Who Becomes an Inventor in America? The Importance of Exposure to Innovation," IDEAS Working Paper Series from RePEc, 2017, https://search.proquest.com/docview/2059036828?pq-origsite=primo; Jonathan Duvall, Sivashankar Sivakanthan, Brandon Daveler, Andrea S. Sundaram, and Rory A. Cooper, "Inventors with Disabilities—an Opportunity for Innovation, Inclusion, and Economic Development," *Technology and Innovation* 22, no. 3 (2022): 315–329, https://doi.org/10.21300

/22.3.2022.5; Holly Fechner and Matthew S. Shapanka, "Closing Diversity Gaps in Innovation: Gender, Race, and Income Disparities in Patenting and Commercialization of Inventions," *Technology & Innovation* 19, no. 4 (2018): 727–734, https://doi.org/10.21300/19.4.2018.727.

Chapter 6

1. Paola Cecchi-Dimeglio and Kim Kleman, "How to React to a Biased Performance Review," in *HBR Guide for Women at Work*, ed. Harvard Business Review, HBR Guide Series (Brighton, MA: Harvard Business Review Press, 2018), 165–173; Derrick P. Bransby, Michaela Kerrissey, and Amy C. Edmondson, "Paradise Lost (and Restored?): A Study of Psychological Safety over Time," *Academy of Management Discoveries* (2024), https://doi.org/10.5465/amd.2023.0084.

2. Paola Cecchi-Dimeglio, *Diversity Dividend: The Transformational Power of Small Changes to Debias Your Company, Attract Diverse Talent, Manage Everyone Better—and Make More Money* (Cambridge, MA: MIT Press, 2023).

3. Andrew Ross Sorkin et al., "How Apple Used Its Car Project to Drive Wider Innovation," DealBook Newsletter, *New York Times*, February 28, 2024, https://www.nytimes.com/2024/02/28/business/dealbook/apple-car-project-to-drive-wider-innovation.html; Mark Gurman and Drake Bennett, "Apple Car's Crash: Design Details, Tim Cook's Indecision, Failed Tesla Deal," *Bloomberg*, March 6, 2024, https://www.bloomberg.com/news/features/2024-03-06/apple-car-s-crash-design-details-tim-cook-s-indecision-failed-tesla-deal.

4. Christine Moy and Adit Gadgil, *Opportunities in the Metaverse* (JPMorgan Chase, 2022), https://www.jpmorgan.com/content/dam/jpm/treasury-services/documents/opportunities-in-the-metaverse.pdf; Nancy Harhut, *Using Behavioral Science in Marketing: Drive Customer Action and Loyalty by Prompting Instinctive Responses* (New York: Kogan Page, 2022).

5. Ilaria Mancuso, Antonio Messeni Petruzzelli, and Umberto Panniello, "Digital Business Model Innovation in Metaverse: How to Approach Virtual Economy Opportunities," *Information Processing & Management* 60, no. 5 (2023): 103457, https://doi.org/10.1016/j.ipm.2023.103457.

6. Scott Brown, "Metaverse in Healthcare," *Accenture* (blog), 2022, https://web.archive.org/web/20230306153413/https://www.accenture.com/us-en/blogs/insight-driven-health/healthcare-in-the-metaverse.

7. Bernard Marr, "The Amazing Possibilities of Healthcare in the Metaverse," *Forbes*, February 25, 2022, https://www.forbes.com/sites/bernardmarr/2022/02/23/the-amazing-possibilities-of-healthcare-in-the-metaverse/; Mancuso et al., "Digital Business Model Innovation in Metaverse."

8. Ozgur Adigozel, Tibor Mérey, and Madeline Mathews, "The Health Care Metaverse Is More Than a Virtual Reality," BCG Global, January 19, 2023, https://www.bcg.com/publications/2023/reaping-the-benefits-of-the-healthcare-metaverse.

9. World Health Organization, "Health Equity," 2024, https://www.who.int/health-topics/health-equity; Neelima Dwivedi and Ruma Bhargawa, "Public–Private Partnerships Can Benefit Healthcare in India," World Economic Forum, September 9, 2022, https://www.weforum.org/agenda/2022/09/public-private-partnerships-india-healthcare-ecosystem/.

10. Marc Carrel-Billiard, Dan Guenther, Nicola Rosa, and Krista Taylor, *Meeting the New Reality with Immersive Learning* (Accenture, 2021), https://www.accenture.com/us-en/insights/technology/immersive-learning.

11. "VR Onboarding: How Accenture Is Redefining HR with Metaverse Technology," *HRD Connect*, 2023, https://www.hrdconnect.com/casestudy/how-accentures-enterprise-metaverse-has-elevated-employee-onboarding/.

12. "How Takeda Harnessed the Power of the Metaverse for Positive Human Impact," Ernst & Young Consulting, 2023, https://www.ey.com/en_us/metaverse-new-ways-interacting-businesses-consumers/how-takeda-harnessed-the-power-of-the-metaverse-for-positive-human-impact.

13. "Microsoft Launches Virtual Worlds–Based Workplace," *E&T*, January 25, 2024, https://eandt.theiet.org/2024/01/25/microsoft-launches-virtual-worlds-based-workplace.

14. Torsten Fell, "Microsoft Integrates Mesh into Teams for Enhanced Virtual Collaboration," Immersive Learning News, February 7, 2024, https://www.immersivelearning.news/2024/02/07/microsoft-integrates-mesh-into-teams-for-enhanced-virtual-collaboration/; David Biello, "How Science Stopped BP's Gulf of Mexico Oil Spill," *Scientific American*, April 19, 2011, https://www.scientificamerican.com/article/how-science-stopped-bp-gulf-of-mexico-oil-spill/.

15. "Global Mercy," Mercy Ships, accessed February 10, 2025, https://mercyships.ca/en/about-us/our-fleet/global-mercy/; Mariska Buitendijk, "All You Want to Know About Mercy Ships' New Hospital Ship Global Mercy," *SWZ Maritime*, July 28, 2022, https://swzmaritime.nl/news/2022/07/28/all-you-want-to-know-about-mercy-ships-new-hospital-ship-global-mercy/.

16. Ramana Nanda and Jesper B. Sørensen, "Workplace Peers and Entrepreneurship," *Management Science* 56, no. 7 (2010): 1116–1126, https://doi.org/10.1287/mnsc.1100.1179.

17. Samuel Awuni Azinga, Anthony Frank Obeng, Florence Y. A. Ellis, and Martin Owusu Ansah, "Assessing the Effects of Transformational Leadership on Innovative Behavior: The Role of Affective Commitment and Psychological Capital,"

Evidence-Based HRM: A Global Forum for Empirical Scholarship 11, no. 4 (2023): 725–745, https://doi.org/10.1108/EBHRM-05-2022-0119; Tahseen Arshi and Venkoba Rao, "Assessing Impact of Employee Engagement on Innovation and the Mediating Role of Readiness for Innovation," *International Journal of Comparative Management* 2, no. 2 (2019): 174–202, https://doi.org/10.1504/IJCM.2019.100857.

18. Viola Isabel Nyssen Guillén and Carsten Deckert, "Cultural Influence on Innovativeness—Links between 'The Culture Map' and the 'Global Innovation Index,'" *International Journal of Corporate Social Responsibility* 6, no. 1 (2021): 1–12, https://doi.org/10.1186/s40991-021-00061-x; David N. Laband and Bernard F. Lentz, "Like Father, like Son: Toward an Economic Theory of Occupational Following," *Southern Economic Journal* 50, no. 2 (1983): 474–493, https://doi.org/10.2307/1058220.

19. Antonio J. Marques et al., "Impact of a Virtual Reality-Based Simulation on Empathy and Attitudes Toward Schizophrenia," *Frontiers in Psychology* 13 (2022): 814984, https://doi.org/10.3389/fpsyg.2022.814984; Dong Wang, Weiqiang Zhang, Shujie Hu, Shoutong Qin, and Dalei Li, "The Work Process Simulation for Pentagon Impact Compactor Based on Virtual Reality Technology," in *2009 IEEE 10th International Conference on Computer-Aided Industrial Design & Conceptual Design* (IEEE, 2009), 874–878, https://doi.org/10.1109/CAIDCD.2009.5375101; Jane Lu, "How to Address the Diversity Challenges of the Metaverse," World Economic Forum, June 14, 2022, https://www.weforum.org/agenda/2022/06/metaverse-platforms-face-diversity-equity-and-inclusion-challenges-heres-how-to-address-them/; "How Virtual Reality (VR) Headsets Help Visually Impaired People Regain Vision and Transform the Way They See the World," *Vision Buddy* (blog), August 21, 2020, https://payments.visionbuddy.com/blogs/the-vision-buddy-blog/how-virtual-reality-vr-headsets-help-visually-impaired-people.

20. Heather, "U.S. FDA Fasttracks Floreo's Virtual Reality System with Admission into Both Breakthrough Device Program and TAP Program," *Floreo Blog*, December 12, 2023, https://floreovr.com/learning-center/blog/us-fda-fasttracks-floreos-virtual-reality-system-with-admission-into-both-breakthrough-device-program-and-tap-program; "Vision Buddy | Give Your Vision a Second Chance," Vision Buddy, 2024, https://visionbuddy.com/.

21. Alex Bell, Raj Chetty, Xavier Jaravel, Neviana Petkova, and John Van Reenen, "Who Becomes an Inventor in America? The Importance of Exposure to Innovation," IDEAS Working Paper Series from RePEc, 2017, https://search.proquest.com/docview/2059036828?pq-origsite=primo; Otto Toivanen and Lotta Väänänen, "Returns to Inventors," *Review of Economics and Statistics* 94, no. 4 (2012): 1173–1190, https://doi.org/10.1162/REST_a_00269.

22. Philippe Aghion, Ufuk Akcigit, Ari Hyytinen, and Otto Toivanen, "The Social Origins of Inventors," preprint, NBER Working Paper Series, December 2017, https://doi.org/10.3386/w24110.

23. "NVIDIA Omniverse," Nvidia, accessed March 7, 2024, https://www.nvidia.com/en-us/omniverse/.

24. Mike Geyer, "Celebrating the Opening of the World's First Virtual Factory in Omniverse," *Nvidia* (blog), March 21, 2023, https://blogs.nvidia.com/blog/bmw-group-nvidia-omniverse/.

25. Ralph Katz, ed., *The Human Side of Managing Technological Innovation: A Collection of Readings* (New York: Oxford University Press, 1997).

26. David Aaker, "The Genius Bar—Branding the Innovation," *Harvard Business Review*, January 5, 2012, https://hbr.org/2012/01/the-genius-bar-branding-the-in; Katie Hafner, "They're Off to See the Wizards," *New York Times*, January 27, 2005, https://www.nytimes.com/2005/01/27/technology/circuits/theyre-off-to-see-the-wizards.html.

27. P. Ranganath Nayak and John Ketteringham, "3M's Post-It Notes: A Managed or Accidental Innovation?," in *The Human Side of Managing Technological Innovation: A Collection of Readings*, ed. Ralph Katz, 2nd ed. (New York: Oxford University Press, 2003), 1–11.

28. Brenda K. Tsai, "Building an Inclusive Metaverse Starts Now. Here's How," World Economic Forum, January 20, 2022, https://www.weforum.org/agenda/2022/01/building-inclusive-metaverse-must-start-now/.

29. Adario Strange, "What Parents Need to Know About the Coming Metaverse," World Economic Forum, November 16, 2021, https://www.weforum.org/agenda/2021/11/parents-metaverse-children-risks-vr-ar/.

30. Kate MacArthur, "Here's How Openness and Inclusion Can Help Scale Immersive Media," World Economic Forum, September 14, 2022, https://www.weforum.org/agenda/2022/09/openness-inclusion-scale-immersive-media/.

31. Malissa Alinor, "Research: The Real-Time Impact of Microaggressions," *Harvard Business Review*, May 17, 2022, https://hbr.org/2022/05/research-the-real-time-impact-of-microaggressions.

32. Sebastian Leape, Jinchen Zou, Olivia Loadwick, Robin Nuttall, Matt Stone, and Bruce Simpson, "The 5Ps of Company Purpose Are Much More than a Mission Statement," McKinsey, November 5, 2020, https://www.mckinsey.com/capabilities/strategy-and-corporate-finance/our-insights/more-than-a-mission-statement-how-the-5ps-embed-purpose-to-deliver-value.

33. Daniela Gutermann, Nale Lehmann-Willenbrock, Diana Boer, Marise Born, and Sven C. Voelpel, "How Leaders Affect Followers' Work Engagement and Performance: Integrating Leader–Member Exchange and Crossover Theory," *British Journal of Management* 28, no. 2 (2017): 299–314, https://doi.org/10.1111/1467-8551.12214.

34. Sara Reardon, "Gender Gap in US Patents Leads to Few Inventions That Help Women," *Nature* 597, no. 7874 (2021): 139–140, https://doi.org/10.1038/d41586-021-02298-9; Elizabeth Wasserman and Evan Sully, "'Next Big Thing' at Risk as Fewer Women, Minorities Get Patents," *Bloomberg Law*, November 25, 2019, https://news.bloomberglaw.com/ip-law/next-big-thing-at-risk-as-fewer-women-minorities-get-patents.

35. Lindsay G. Oades, Michael Steger, Antonella Delle Fave, and Jonathan Passmore, eds., *The Wiley Blackwell Handbook of the Psychology of Positivity and Strengths-Based Approaches at Work*, Wiley Blackwell Handbooks in Organizational Psychology (Hoboken, NJ: Wiley-Blackwell, 2016), https://doi.org/10.1002/9781118977620.

36. Kim A. Johnston and Maureen Taylor, eds., *The Handbook of Communication Engagement*, Handbooks in Communication and Media (Hoboken, NJ: Wiley-Blackwell, 2018), https://doi.org/10.1002/9781119167600.

37. Paola Cecchi-Dimeglio, Taha Masood, and Andy Ouderkirk, "What Makes Innovation Partnerships Succeed," *Harvard Business Review*, July 14, 2022, https://hbr.org/2022/07/what-makes-innovation-partnerships-succeed.

38. Gutermann et al., "How Leaders Affect Followers' Work Engagement and Performance."

39. Catherine H. Crouch and Eric Mazur, "Peer Instruction: Ten Years of Experience and Results," *American Journal of Physics* 69, no. 9 (2001): 970–977, https://doi.org/10.1119/1.1374249.

40. Kelly Palmer and David Blake, "How to Help Your Employees Learn from Each Other," *Harvard Business Review*, November 8, 2018, https://hbr.org/2018/11/how-to-help-your-employees-learn-from-each-other; R. Keith Sawyer, "The Cambridge Handbook of the Learning Sciences," in *The Cambridge Handbook of the Learning Sciences*, ed. R. Keith Sawyer (Cambridge: Cambridge University Press, 2014), iii, https://doi.org/10.1017/CBO9781139519526.

Chapter 7

1. National Research Council, *Continuing Innovation in Information Technology* (Washington, DC: National Academies Press, 2012), https://doi.org/10.17226/13427.

2. "Meta Announces Tools and Partnerships to Bring VR Productivity and Collaboration to Businesses," Auganix, October 11, 2022, https://www.auganix.org/meta-announces-tools-and-partnerships-to-bring-vr-productivity-and-collaboration-to-businesses/; "Meta Strikes a Partnership with Google to Include Its Search Results in Assistant's Responses," CNBC, April 18, 2024, https://www.cnbc.com/video/2024/04/18/meta-strikes-a-partnership-with-google-to-include-its-search-results-in-assistants-responses.html.

3. "Galaxy Book4 Series + Windows 11 Pro—the Platform for Business," *Insights*, April 4, 2024, https://insights.samsung.com/2024/04/04/galaxy-book4-windows-11-pro-the-platform-for-business/.

4. Stefana Raileanu and Ramya Ravi, "Accelerate Your AI Workloads with Intel® Optimized AI Software on Amazon Web Services (AWS)*," Intel, accessed February 10, 2025, https://www.intel.com/content/www/us/en/developer/articles/technical/accelerate-your-ai-software-workloads-on-aws.html.

5. "IBM Closes Landmark Acquisition of Red Hat for $34 Billion; Defines Open, Hybrid Cloud Future," Red Hat, July 9, 2019, https://www.redhat.com/en/about/press-releases/ibm-closes-landmark-acquisition-red-hat-34-billion-defines-open-hybrid-cloud-future.

6. "The Data Center Will Never Be the Same: Modernizing the Data Center with VMware and NVIDIA," Nvidia, accessed February 10, 2025, https://www.nvidia.com/en-us/data-center/vmware/.

7. Jeff Teper, "Microsoft and Meta Partner to Deliver Immersive Experiences for the Future of Work and Play," *Microsoft* (blog), October 11, 2022, https://blogs.microsoft.com/blog/2022/10/11/microsoft-and-meta-partner-to-deliver-immersive-experiences-for-the-future-of-work-and-play/.

8. "Sony and KIRKBI Invest in Epic Games to Build the Future of Digital Entertainment," Epic Games, April 11, 2022, https://www.epicgames.com/site/en-US/news/sony-and-kirkbi-invest-in-epic-games-to-build-the-future-of-digital-entertainment.

9. Chase DiBenedetto, "Google Launches New Gaming World for Kids on Roblox," *Mashable*, September 26, 2024, https://mashable.com/article/new-roblox-game-for-kids-teaches-internet-safety; "Hewlett Packard Enterprise and Microsoft Announce Plans to Deliver Integrated Hybrid IT Infrastructure," Microsoft, December 1, 2015, https://news.microsoft.com/2015/12/01/hewlett-packard-enterprise-and-microsoft-announce-plans-to-deliver-integrated-hybrid-it-infrastructure/.

10. Aaron Back, "Microsoft & FedEx Partnership: Commerce Redefined with Logistics-as-a-Service," *Cloud Wars*, January 26, 2022, https://cloudwars.com/ai/microsoft-fedex-partnership-commerce-redefined-with-logistics-as-a-service/.

11. Stefano Mazzocchi, "Open Innovation: The New Imperative for Creating and Profiting from Technology," *Innovation* 6, no. 3 (2004): 474, https://doi.org/10.5172/impp.2004.6.3.474; Walter W. Powell, Kenneth W. Koput, and Laurel Smith-Doerr, "Interorganizational Collaboration and the Locus of Innovation: Networks of Learning in Biotechnology," *Administrative Science Quarterly* 41, no. 1 (1996): 116–145, https://doi.org/10.2307/2393988.

12. Paola Cecchi-Dimeglio, Taha Masood, and Andy Ouderkirk, "What Makes Innovation Partnerships Succeed," *Harvard Business Review*, July 14, 2022, https://hbr.org/2022/07/what-makes-innovation-partnerships-succeed.

13. Nicholas Carr, *The Big Switch: Rewiring the World, from Edison to Google*, rep. ed. (W. W. Norton, 2013); Daisuke Wakabayashi and Jack Nicas, "Apple, Google and a Deal That Controls the Internet," *New York Times*, October 25, 2020, https://www.nytimes.com/2020/10/25/technology/apple-google-search-antitrust.html.

14. Peter F. Drucker, *Managing for the Future: The 1990s and Beyond* (New York: Dutton, 1992).

15. David J. Teece, Gary Pisano, and Amy Shuen, "Dynamic Capabilities and Strategic Management," *Strategic Management Journal* 18, no. 7 (1998): 509–533, https://doi.org/10.1002/(SICI)1097-0266(199708)18:7<509::AID-SMJ882>3.0.CO;2-Z.

16. Joe Tidd and John Bessant, *Managing Innovation: Integrating Technological, Market and Organizational Change*, 7th ed. (Hoboken, NJ: Wiley, 2021).

17. Eduardo Salas, Ramón Rico, and Jonathan Passmore, eds., *The Wiley Blackwell Handbook of the Psychology of Team Working and Collaborative Processes*, Wiley Blackwell Handbooks in Organizational Psychology (Hoboken, NJ: Wiley-Blackwell, 2017), https://doi.org/10.1002/9781118909997.

18. Michael D. Watkins, "Why DaimlerChrysler Never Got into Gear," *Harvard Business Review*, May 18, 2007, https://hbr.org/2007/05/why-the-daimlerchrysler-merger.

19. Svenolof Karlsson and Anders Lugn, "Three Important Decisions," Ericsson, accessed February 10, 2025, https://www.ericsson.com/en/about-us/history/changing-the-world/big-bang/three-important-decisions; Puneet Pal Singh, "Can Sony Succeed Where Sony–Ericsson Partnership Failed?," BBC, October 13, 2011, https://www.bbc.com/news/business-15285258.

20. Tom Warren, "Microsoft Wasted at Least $8 Billion on Its Failed Nokia Experiment," *The Verge*, May 25, 2016, https://www.theverge.com/2016/5/25/11766540/microsoft-nokia-acquisition-costs.

21. Peter W. G. Morris, Jeffrey K. Pinto, and Jonas Söderlund, eds., *The Oxford Handbook of Project Management* (Oxford: Oxford University Press, 2011), https://doi.org/10.1093/oxfordhb/9780199563142.001.0001.

22. Jeffrey J. Reuer, ed., *Strategic Alliances: Theory and Evidence*, Oxford Management Readers (Oxford: Oxford University Press, 2004); Jeffrey H. Dyer and Harbir Singh, "The Relational View: Cooperative Strategy and Sources of Interorganizational Competitive Advantage," *Academy of Management Review* 23, no. 4 (1998): 660–679, https://doi.org/10.2307/259056.

23. Charlott Hübel, Ilka Weissbrod, and Stefan Schaltegger, "Strategic Alliances for Corporate Sustainability Innovation: The 'How' and 'When' of Learning Processes," *Long Range Planning* 55, no. 6 (2022): 102200, https://doi.org/10.1016/j.lrp.2022.102200.

24. Bent Flyvbjerg, ed., *The Oxford Handbook of Megaproject Management* (Oxford: Oxford University Press, 2017), https://doi.org/10.1093/oxfordhb/9780198732242.001.0001.

25. Peter Kamminga, "Rethinking Contract Design: Why Incorporating Non-Legal Drivers of Contractual Behavior in Contracts May Lead to Better Results in Complex Defense Systems Procurement," *Journal of Public Procurement* 15, no. 2 (2015): 208–235, https://doi.org/10.1108/JOPP-15-02-2015-B004.

26. Paola Cecchi-Dimeglio, Taha Masood, and Andy Ouderkirk, "What Makes Innovation Partnerships Succeed," *Harvard Business Review*, July 14, 2022, https://hbr.org/2022/07/what-makes-innovation-partnerships-succeed.

27. Elie Ofek, Eitan Muller, and Barak Libai, *Innovation Equity: Assessing and Managing the Monetary Value of New Products and Services* (Chicago: University of Chicago Press, 2016).

28. Sunil Chahal, "Agile Methodologies for Improved Product Management," *Journal of Business and Strategic Management* 8, no. 4 (2023): 79–94, https://doi.org/10.47941/jbsm.1439.

29. Bernd H. Schmitt, *Customer Experience Management: A Revolutionary Approach to Connecting with Your Customers* (Hoboken, NJ: John Wiley & Sons, 2003), https://archive.org/details/customerexperien0000schm/page/n2/mode/1up.

30. Hua Hsu, "The Year in 'Diversity Fatigue,'" *New Yorker*, December 26, 2017, https://www.newyorker.com/culture/2017-in-review/the-year-in-diversity-fatigue; Gabi Thesing, "Do You Have 'Diversity Fatigue'? Here Are 5 Ways for Businesses to Get Out of Their DEI Slump," World Economic Forum, April 11, 2023, https://www.weforum.org/agenda/2023/04/how-businesses-can-beat-diversity-fatigue-dei/.

31. Paola Cecchi-Dimeglio, "Ask Dr Paola: How AI Is Changing the Legal Industry," *Thomson Reuters Legal Solutions* (blog), February 7, 2018, https://legalsolutions.thomsonreuters.co.uk/blog/2018/02/07/ask-dr-paola-ai-changing-legal-industry/.

32. "Consumer Interest in 'Virtual Living' Intensifies, Accenture Survey Finds," Accenture Newsroom, April 27, 2022, https://newsroom.accenture.com/news/2022/consumer-interest-in-virtual-living-intensifies-accenture-survey-finds; Florian Zandt, "How Popular Are VR and AR Headsets?," Statista, January 19, 2024, https://www.statista.com/chart/31599/vr-ar-headset-excitement/.

33. Cade Metz, "'The Godfather of A.I.' Leaves Google and Warns of Danger Ahead," *New York Times*, May 1, 2023, https://www.nytimes.com/2023/05/01/technology/ai-google-chatbot-engineer-quits-hinton.html.

34. Nich Richie, "What the VRChat Controversy Teaches Us About Accessibility in Games," GamesHub, October 21, 2022, https://www.gameshub.com/news/opinions-analysis/what-the-vrchat-controversy-teaches-us-about-accessibility-in-games-31936/.

35. Alison Jane Martingano, Ja-Nae Duane, Ellenor Brown, and Susan Persky, "Demographic Differences in Presence Across Seven Studies," *Virtual Reality* 27, no. 3 (2023): 2297–2313, https://doi.org/10.1007/s10055-023-00805-z.

36. Yogesh K. Dwivedi et al., "Metaverse Marketing: How the Metaverse Will Shape the Future of Consumer Research and Practice," *Psychology & Marketing* 40, no. 4 (2023): 750–776, https://doi.org/10.1002/mar.21767.

37. Ko de Ruyter, Jonas Heller, Tim Hilken, Matthew Chylinski, Debbie I. Keeling, and Dominik Mahr, "Seeing with the Customer's Eye: Exploring the Challenges and Opportunities of AR Advertising," *Journal of Advertising* 49, no. 2 (2020): 109–124, https://doi.org/10.1080/00913367.2020.1740123.

38. Tim Hilken, Mathew Chylinski, Debbie I. Keeling, Jonas Heller, Ko de Ruyter, and Dominik Mahr, "How to Strategically Choose or Combine Augmented and Virtual Reality for Improved Online Experiential Retailing," *Psychology & Marketing* 39, no. 3 (2022): 495–507, https://doi.org/10.1002/mar.21600; Maja Golf-Papez et al., "Embracing Falsity Through the Metaverse: The Case of Synthetic Customer Experiences," *Business Horizons* 65, no. 6 (2022): 739–749, https://doi.org/10.1016/j.bushor.2022.07.007.

39. Thomas Alsop, "AR Glasses and Headsets—Statistics and Facts," Statista, 2023, https://www.statista.com/topics/10134/ar-glasses/.

40. Thomas Alsop, "Consumer Augmented Reality (AR) Glasses Unit Sales Worldwide from 2019 to 2024 [Graph]," Statista, December 14, 2022, https://www.statista.com/statistics/1221567/consumer-ar-glasses-unitsales-worldwide/; "Vision Pro," Apple, 2023, https://www.apple.com/apple-vision-pro/; Lauren Goode and Michael Calore, "Meta Announces New Quest 3 Headset," *Wired*, June 1, 2023, https://www.wired.com/story/meta-quest-3-vr-headset-price-specs-release-date/.

41. Allen Firstenberg and Jason Salas, *Designing and Developing for Google Glass: Thinking Differently for a New Platform* (Sebastopol, CA: O'Reilly Media, 2015); Ben D. Sawyer, Victor S. Finomore, Andres A. Calvo, and P. A. Hancock, "Google Glass: A Driver Distraction Cause or Cure?," *Human Factors* 56, no. 7 (2014): 1307–1321, https://doi.org/10.1177/0018720814555723.

42. Jibo He, William Choi, Jason S. McCarley, Barbara S. Chaparro, and Chun Wang, "Texting While Driving Using Google Glass™: Promising but Not Distraction-Free," *Accident Analysis & Prevention* 81 (2015): 218–229, https://doi.org/10.1016/j.aap.2015.03.033.

43. Daniel Harley, "The Promise of Beginnings: Unpacking 'Diversity' at Oculus VR," *Convergence: The International Journal of Research into New Media Technologies* 29, no. 2 (2023): 417–431, https://doi.org/10.1177/13548565221122911.

44. Wikipedia, "Quest 2," last modified January 1, 2025 4:12 (UTC), https://en.wikipedia.org/wiki/Quest_2.

45. Paola Cecchi-Dimeglio, "How Leaders Harness the Extraordinary Power of Uniqueness," *Forbes*, October 26, 2023, https://www.forbes.com/sites/paolacecchi-dimeglio/2023/10/26/how-leaders-harness-the-extraordinary-power-of-uniqueness/.

Chapter 8

1. Tanveer Nayak, Saransh Kalambele, and Anurag Jain, "Metaverse in Public Sector," in *The Business of the Metaverse*, ed. Hemachandran K. and Raul V. Rodriguez (New York: Productivity Press, 2023), 259–266, https://doi.org/10.4324/b23404-17; Rajan Gupta and Saibal K. Pal, "Metaverse for Public Sector," in *Introduction to Metaverse: Technology Landscape, Applications, and Challenges* (Singapore: Springer Singapore, 2023), 91–114, https://doi.org/10.1007/978-981-99-7397-2_5; Cathy Li, "Who Will Govern the Metaverse?," World Economic Forum, May 25, 2022, https://www.weforum.org/agenda/2022/05/metaverse-governance/.

2. Robert Tross, James Chung, Tahir Bangash, Parker Lytle, Devon Halley, and Lauren Nishikawa, "The Traveler's Guide to Unlocking the Value of Metaverse in Government," *Deloitte Insights*, June 1, 2023, https://www2.deloitte.com/us/en/insights/industry/public-sector/metaverse-technologies-public-and-government.html.

3. Decerry Donato, "Santa Monica Is Using the Metaverse to Gamify Its Shopping District," *Dot.la*, December 13, 2021, https://dot.la/santa-monica-metaverse-2656021933.html.

4. Aaron Limbu, "Hong Kong's MTR Becomes World's 1st Transport Operator to Enter The Sandbox Metaverse," *BlockchainNews*, April 22, 2022, https://blockchain.news/news/hong-kongs-mtr-becomes-worlds-1st-transport-operator-to-enter-the-sandbox-metaverse.

5. SMG, "Seoul, First Local Gov't to Start New-Concept Public Service with 'Metaverse Platform,'" Seoul Metropolitan Government, November 8, 2021, https://english.seoul.go.kr/seoul-first-local-govt-to-start-new-concept-public-service-with-metaverse-platform/.

6. "How the Republic of Korea Became a World ICT Leader?," ITU, May 29, 2020, https://www.itu.int/hub/2020/05/how-the-republic-of-korea-became-a-world-ict-leader/.

7. SMG, "Seoul, First Local Gov't to Start New-Concept Public Service with 'Metaverse Platform.'"

8. Jim Wyss, "Barbados Is Opening a Diplomatic Embassy in the Metaverse," *Bloomberg*, December 14, 2021, https://www.bnnbloomberg.ca/barbados-is-opening-a-diplomatic-embassy-in-the-metaverse-1.1695625.

9. Katerina Girginova, "Global Visions for a Metaverse," *International Journal of Cultural Studies* (2024), https://doi.org/10.1177/13678779231224799; Erman Akilli,

"The Metaverse Diplomacy," *Insight Turkey* 24, no. 3 (2022): 67–88, https://doi.org/10.25253/99.2022243.6.

10. Anand Singh Rajawat, S. B. Goyal, RamKumar Solanki, Maria Simona Raboaca, Traian Candin Mihaltan, and Zoltán Illés, "Blockchain-Based Security Framework for Metaverse: A Decentralized Approach," in *2023 15th International Conference on Electronics, Computers and Artificial Intelligence (ECAI)* (IEEE, 2023), 1–6, https://doi.org/10.1109/ECAI58194.2023.10193962; Winston Ma and Ken Huang, *Blockchain and Web3: Building the Cryptocurrency, Privacy, and Security Foundations of the Metaverse* (Hoboken, NJ: Wiley, 2022).

11. Brian Prince, "South Carolina Hit in Massive Cyberattack—3.6 Million Tax Payers Exposed," *SecurityWeek*, October 26, 2012, https://www.securityweek.com/south-carolina-hit-massive-cyberattack/; Jim Forsyth, "Private Records of 3.5 Million People Exposed by Texas," *Reuters*, April 11, 2011, https://www.reuters.com/article/idUSTRE73A5MF/.

12. Jeremiah Rozman, *The Synthetic Training Environment* (Arlington, VA: Association of the United States Army, 2020), https://www.ausa.org/sites/default/files/publications/SL-20-6-The-Synthetic-Training-Environment.pdf.

13. Mandy Mayfield, "Virtual, Augmented Reality Tech Transforming Training," *National Defense*, February 17, 2021, https://www.nationaldefensemagazine.org/articles/2021/2/17/virtual-augmented-reality-tech-transforming-training; Katie Lange, "Virtual, Augmented Reality Are Moving Warfighting Forward," US Department of Defense, February 10, 2020, https://www.defense.gov/News/Feature-Stories/Story/Article/2079205/virtual-augmented-reality-are-moving-warfighting-forward/.

14. APIs (application programming interfaces) enable different applications, platforms, and services to communicate efficiently, allowing users to move data and digital assets across various metaverse environments without friction. Meanwhile, microservice architectures break applications into modular, independent components, making systems more scalable, more flexible, and easier to update. This approach reduces downtime, enhances performance, and supports continuous innovation—critical factors in a rapidly evolving digital ecosystem like the metaverse. By leveraging these technologies, governments and public sector organizations can ensure their applications remain adaptable, user-friendly, and compatible with emerging platforms.

15. Roberto Di Pietro and Stefano Cresci, "Metaverse: Security and Privacy Issues," in *2021 Third IEEE International Conference on Trust, Privacy and Security in Intelligent Systems and Applications (TPS-ISA)* (IEEE, 2021), 281–288, https://doi.org/10.1109/TPSISA52974.2021.00032; Yan Huang, Yi Joy Li, and Zhipeng Cai, "Security and Privacy in Metaverse: A Comprehensive Survey," *Big Data Mining and Analytics* 6, no. 2 (2023): 234–247, https://doi.org/10.26599/BDMA.2022.9020047.

16. *Privacy and Safety in the Metaverse* (World Economic Forum, July 12, 2023), https://www.weforum.org/publications/privacy-and-safety-in-the-metaverse/.

17. "Education 4.0," World Economic Forum, accessed February 11, 2025, https://initiatives.weforum.org/reskilling-revolution/education-4-0 and https://widgets.weforum.org/reskillingrevolution/initiatives/forum-led/education-4-0/index.html?

18. Tim Levin, "AI Can Make Anyone a Programmer and Has 'Closed the Digital Divide,' Nvidia CEO Says," *Business Insider*, May 29, 2023, https://www.businessinsider.com/ai-tools-turn-anyone-into-programmer-nvidia-ceo-2023-5; Joe Tidd and John Bessant, *Managing Innovation: Integrating Technological, Market and Organizational Change*, 7th ed. (Hoboken, NJ: Wiley, 2021); Kathy Hirsch-Pasek et al., "A Whole New World: Education Meets the Metaverse," Brookings, February 14, 2022, https://www.brookings.edu/research/a-whole-new-world-education-meets-the-Metaverse/; Victoria Masterton, "Future of Jobs 2023: These Are the Most In-Demand Skills Now—and Beyond," World Economic Forum, May 1, 2023, https://www.weforum.org/stories/2023/05/future-of-jobs-2023-skills/#:~:text=The%20World%20Economic%20Forum's%20Future,expected%20to%20see%20growing%20demand.

19. See https://www.whosemetaverse.org.

20. Mads T. Bonde, Guido Makransky, Jakob Wandall, Mette V. Larsen, Mikkel Morsing, Hanne Jarmer, and Morten O. A. Sommer, "Improving Biotech Education Through Gamified Laboratory Simulations," *Nature Biotechnology* 32, no. 7 (2014): 694.

21. Scott Likens and Andrea Mower, "What Does Virtual Reality and the Metaverse Mean for Training?," PwC, September 15, 2022, https://www.pwc.com/us/en/tech-effect/emerging-tech/virtual-reality-study.html.

22. Adam Hadhazy, "Stanford Course Allows Students to Learn About Virtual Reality While Fully Immersed in VR Environments," *Stanford Report*, November 5, 2021, https://news.stanford.edu/2021/11/05/new-class-among-first-taught-entirely-virtual-reality/.

23. Hirsch-Pasek et al., "A Whole New World."

24. Iván Sánchez-López, Rosabel Roig-Vila, and Amor Pérez-Rodríguez, "Metaverse and Education: The Pioneering Case of Minecraft in Immersive Digital Learning," *El profesional de la información* 31, no. 6 (2022), https://doi.org/10.3145/epi.2022.nov.10.

25. Lee Waller and Sharon Waller, eds., *Higher Education: Reflections from the Field*, vol. 3, Education and Human Development Series (London: IntechOpen, 2023), https://doi.org/10.5772/intechopen.112127.

26. Omar Sadek, Fiona Baldwin, Rebecca Gray, Nadine Khayyat, and Theofanis Fotis, "Impact of Virtual and Augmented Reality on Quality of Medical Education During the COVID-19 Pandemic: A Systematic Review," *Journal of Graduate Medical Education* 15, no. 3 (2023): 328–338, https://doi.org/10.4300/JGME-D-22-00594.1.

27. Xieling Chen, Di Zou, Haoran Xie, and Fu Lee Wang, "Metaverse in Education: Contributors, Cooperations, and Research Themes," *IEEE Transactions on Learning Technologies* 16, no. 6 (2023): 1111–1129, https://ieeexplore.ieee.org/abstract/document/10129849/; Eugy Han, Mark Roman Miller, Nilam Ram, Kristine L. Nowak, and Jeremy N. Bailenson, "Understanding Group Behavior in Virtual Reality: A Large-Scale, Longitudinal Study in the Metaverse," paper presented at the 72nd Annual International Communication Association Conference, Paris, France, May 18, 2022, https://papers.ssrn.com/sol3/papers.cfm?abstract_id=4110154.

28. Emin Ibili et al., "Investigation of Learners' Behavioral Intentions to Use Metaverse Learning Environment in Higher Education: A Virtual Computer Laboratory," *Interactive Learning Environments* 32, no. 10 (2023): 5893–5918, https://doi.org/10.1080/10494820.2023.2240860; Maria José Sá and Sandro Serpa, "Metaverse as a Learning Environment: Some Considerations," *Sustainability* 15, no. 3 (2023): 2186, https://doi.org/10.3390/su15032186.

29. Jaziar Radianti, Tim A. Majchrzak, Jennifer Fromm, and Isabell Wohlgenannt, "A Systematic Review of Immersive Virtual Reality Applications for Higher Education: Design Elements, Lessons Learned, and Research Agenda," *Computers & Education* 147 (2020): 103778, https://doi.org/10.1016/j.compedu.2019.103778.

30. Amber Dailey-Hebert, "Student Perspectives on Using Virtual Reality to Create Informal Connection and Engagement," *InSight: A Journal of Scholarly Teaching* 17 (2022): 28–46, https://doi.org/10.46504/17202202da.

31. Luis Alberto Laurens-Arredondo and Lilibeth Laurens, "Metaversity: Beyond Emerging Educational Technology," *Sustainability* 15, no. 22 (2023): 15844, https://doi.org/10.3390/su152215844.

32. Neli Maria Mengalli and Antonio Aparecido Carvalho, "Metaverse in Higher Education and the Metaversities: Disruptive Technologies and Innovations in Industry 5.0 for Phygital Transformation," in *Educational Perspectives on Digital Technologies in Modeling and Management*, ed. G. S. Prakasha, Maria Lapina, Deepanraj Balakrishnan, and Mohammad Sajid (Hershey, PA: IGI Global, 2024), https://doi.org/10.4018/979-8-3693-2314-4.ch011.

33. Sheila Jagannathan, "How Could the Metaverse Impact Education?," World Economic Forum, December 7, 2022, https://www.weforum.org/agenda/2022/12/metaverse-impact-education-learning/.

34. Sanaa Kaddoura and Fatima Al Husseiny, "The Rising Trend of Metaverse in Education: Challenges, Opportunities, and Ethical Considerations," *PeerJ Computer Science* 9 (2023): e1252, https://doi.org/10.7717/peerj-cs.1252.

35. Mahir Pradana and Hanifah Putri Elisa, "Metaverse in Education: A Systematic Literature Review," *Cogent Social Sciences* 9, no. 2 (2023): 2252656, https://doi.org/10.1080/23311886.2023.2252656.

36. Manca Opara and Žiga Kozinc, "Virtual Reality Training for Management of Chronic Neck Pain: A Systematic Review with Meta-Analysis," *European Journal of Physiotherapy* 26, no. 3 (2023): 135–147, https://doi.org/10.1080/21679169.2023.2215831.

37. "Johns Hopkins Performs Its First Augmented Reality Surgeries in Patients," Johns Hopkins Medicine, February 16, 2021, https://www.hopkinsmedicine.org/news/articles/2021/02/johns-hopkins-performs-its-first-augmented-reality-surgeries-in-patients.

38. Scott Brown, "Metaverse in Healthcare," *Accenture* (blog), 2022, https://web.archive.org/web/20230306153413/https://www.accenture.com/us-en/blogs/insight-driven-health/healthcare-in-the-metaverse.

39. Amir H. Sadeghi et al., "Current and Future Applications of Virtual, Augmented, and Mixed Reality in Cardiothoracic Surgery," *Annals of Thoracic Surgery* 113, no. 2 (2022): 681–691, https://doi.org/10.1016/j.athoracsur.2020.11.030.

40. "Simulation Learning Labs," University of Central Florida, College of Nursing, accessed February 11, 2025, https://nursing.ucf.edu/about/simulation-labs-classrooms.

41. "3D Anatomic Modeling Laboratories," Mayo Clinic, https://www.mayoclinic.org/departments-centers/anatomic-modeling-laboratories/overview/ovc-20473121; see also Barton Goldenberg, "Mayo Clinic's Innovations in XR/VR/AR: A Closer Look," ISM, August 3, 2023, https://ismguide.com/mayo-clinic-xr-metaverse-story.

42. "4 Examples of VR Training in Healthcare," VR Owl, accessed February 11, 2025, https://www.vrowl.io/4-examples-of-vr-training-in-healthcare.

43. Brown, "Metaverse in Healthcare."

44. University of Pittsburgh Medical Center (UPMC) is leveraging digital twin technology to enhance patient care by creating virtual replicas for simulating treatments and predicting outcomes. While details on its use in heart surgery training are limited, UPMC's Cardiothoracic Surgery Research Training Program, funded by an NIH T32 grant, supports advancements in surgical techniques. Additionally, the University of Pittsburgh's School of Pharmacy integrates VR and AR in cardiovascular education. See University of Pittsburgh, Department of Cardiothoracic Surgery, https://www.ctsurgery.pitt.edu; Karen Blum, "How to Cover 'Digital Twins' in Health Care," Association of Health Care Journalists, January 8, 2024, https://healthjournalism.org/blog/2024/01/how-to-cover-digital-twins-in-health-care.

45. Bokyung Kye, Nara Han, Eunji Kim, Yeonjeong Park, and Soyoung Jo, "Educational Applications of Metaverse: Possibilities and Limitations," *Journal of Educational Evaluation for Health Professions* 18 (2021): 32, https://doi.org/10.3352/jeehp.2021.18.32.

46. "New VR Platform Fuses Physical and Virtual Worlds in Parkinson's Disease and Beyond," Cleveland Clinic, July 28, 2021, https://consultqd.clevelandclinic.org/new

-vr-platform-fuses-physical-and-virtual-worlds-in-parkinsons-disease-and-beyond; Becky Boban, "Why Cleveland Owns the Future of Virtual Reality," *Cleveland Magazine*, February 26, 2023, https://clevelandmagazine.com/in-the-cle/news/articles/why-cleveland-owns-the-future-of-virtual-reality.

47. "Is Home-Based Treatment the Future of Cancer Care?," Mayo Clinic, January 16, 2025, https://www.mayoclinic.org/medical-professionals/cancer/news/is-home-based-treatment-the-future-of-cancer-care/mac-20577724.

48. Vincent Jacobbi, "Mayo Clinic Research on Remote Hospital Care for a Rural or Urban Home Setting," *Mayo Clinic News Network*, May 8, 2023, https://newsnetwork.mayoclinic.org/discussion/mayo-clinic-research-on-remote-hospital-care-for-a-rural-or-urban-home-setting.

49. Himani Mittal, "Virtual Reality Applications in Healthcare: A New Age Technology Perspective," in *Immersive Virtual and Augmented Reality in Healthcare*, ed. Rajendra Kumar, Vishal Jain, Garry Tan Wei Han, and Adberezak Touzene (Boca Raton, FL: CRC Press, 2023), 50–62, https://doi.org/10.1201/9781003340133-3; Hassan A. Aziz, "Virtual Reality Programs Applications in Healthcare," *Journal of Health & Medical Informatics* 9, no. 1 (2018), https://doi.org/10.4172/2157-7420.1000305.

50. Theodore Oing and Julie Prescott, "Implementations of Virtual Reality for Anxiety-Related Disorders: Systematic Review," *JMIR Serious Games* 6, no. 4 (2018): e10965, https://doi.org/10.2196/10965.

51. Jessica Maxwell, "DOD Using Virtual Reality for PTSD Treatment," US Army, May 3, 2010, https://www.army.mil/article/38464/dod_using_virtual_reality_for_ptsd_treatment.

52. Jameela Al-Jaroodi, Nader Mohamed, and Eman Abukhousa, "Health 4.0: On the Way to Realizing the Healthcare of the Future," *IEEE Access* 8 (2020): 211189–211210, https://doi.org/10.1109/ACCESS.2020.3038858.

Chapter 9

1. Yogesh K. Dwivedi et al., "Metaverse Marketing: How the Metaverse Will Shape the Future of Consumer Research and Practice," *Psychology & Marketing* 40, no. 4 (2023): 750–776, https://doi.org/10.1002/mar.21767; Thorsten Hennig-Thurau and Björn Ognibeni, "Metaverse Marketing," *NIM Marketing Intelligence Review* 14, no. 2 (2022): 43–47, https://doi.org/10.2478/nimmir-2022-0016.

2. Dogan Gursoy, Lu Lu, Robin Nunkoo, and Demi Deng, "Metaverse in Services Marketing: An Overview and Future Research Directions," *Service Industries Journal* 43, no. 15–16 (2023): 1140–1172, https://doi.org/10.1080/02642069.2023.2252750.

3. Daniel Moise, Amelia Diaconu, Mihaela Diana Oancea Negescu, and Mihai Dinu, "Metaverse Marketing—the Future of Sustainable Marketing," *European Journal of*

Sustainable Development 12, no. 4 (2023): 260–266, https://doi.org/10.14207/ejsd.2023.v12n4p260.

4. "Metaverse AR & VR Hardware: Worldwide," Statista, accessed February 27, 2025, https://www.statista.com/outlook/amo/metaverse/metaverse-ar-vr-hardware/worldwide#.

5. Shivani Zoting, "Metaverse Market Size, Share, and Trends 2025 to 2034," Precedence Research, last updated February 21, 2024, https://www.precedenceresearch.com/metaverse-market; Astrid Eira, "193 Technology Statistics You Must Know: 2024 Market Share Analysis & Data," Finances Online, last modified December 25, 2024, https://financesonline.com/technology-statistics/.

6. Eric Hazan, Greg Kelly, Hamza Khan, Dennis Spillecke, and Lareina Yee, "Marketing in the Metaverse: An Opportunity for Innovation and Experimentation," McKinsey, May 24, 2022, https://www.mckinsey.com/capabilities/growth-marketing-and-sales/our-insights/marketing-in-the-metaverse-an-opportunity-for-innovation-and-experimentation.

7. Amy C. Edmondson, *The Fearless Organization: Creating Psychological Safety in the Workplace for Learning, Innovation, and Growth* (Hoboken, NJ: Wiley-Blackwell, 2018).

8. QuHarrison Terry and Scott "DJ Skee" Keeney, *The Metaverse Handbook: Innovating for the Internet's Next Tectonic Shift* (Hoboken, NJ: John Wiley & Sons, 2022).

9. Aisha Malik, "Epic Raises $2B at a Nearly $32B Valuation to Build Its Kid-Friendly Metaverse," *TechCrunch*, April 11, 2022, https://techcrunch.com/2022/04/11/epic-2b-nearly-32b-valuation-build-kid-friendly-metaverse/.

10. Ievgen Gnitetskyi, Serhii Lebedenko, and Olena Lymar, "Marketing in the Metaverse," *Marketing and Digital Technologies* 7, no. 2 (2023): 119–136, https://doi.org/10.15276/mdt.7.2.2023.9.

11. "Vans Launches 'Vans World' Skatepark Experience in the Roblox Metaverse," Roblox, September 1, 2021, https://corp.roblox.com/newsroom/2021/09/vans-launches-vans-world-skatepark-experience-roblox-metaverse.

12. "Gucci Garden on Roblox," Gucci, accessed February 11, 2025, https://www.gucci.com/us/en/st/stories/article/gucci-gaming-roblox.

13. Danny Stefanic, "Explore Virtual Trade Shows and Expos in the Metaverse Today!," *Hyperspace^{mv}—the Metaverse for Business Platform* (blog), December 9, 2023, https://hyperspace.mv/metaverse-trade-shows/.

14. For millennial shopping and metaverse commerce trends, see the combined statistics from Capital One, Accenture, and Khoros: "Millennial Shopping Statistics," Capital One Shopping Research, last updated October 1, 2024, https://capitaloneshopping.com/research/millennial-shopping-statistics; "Growing Consumer and

Business Interest in the Metaverse Expected to Fuel Trillion Dollar Opportunity for Commerce, Accenture Finds," *Accenture Newsroom*, January 4, 2023, https://newsroom.accenture.com/news/2023/growing-consumer-and-business-interest-in-the-metaverse-expected-to-fuel-trillion-dollar-opportunity-for-commerce-accenture-finds; "The 2024 Social Media Demographics Guide," Khoros, accessed January 30, 2025, https://khoros.com/resources/social-media-demographics-guide.

15. Jiajing Wu, Kaixin Lin, Dan Lin, Ziye Zheng, Huawei Huang, and Zibin Zheng, "Financial Crimes in Web3-Empowered Metaverse: Taxonomy, Countermeasures, and Opportunities," *IEEE Open Journal of the Computer Society* 4 (2023): 37–49, https://doi.org/10.1109/OJCS.2023.3245801; Cheng Xu, Xueji Liang, Yanqi Sun, and Xudong He, "Fraudsters Beware: Unleashing the Power of Metaverse Technology to Uncover Financial Fraud," *International Journal of Human–Computer Interaction* 40, no. 18 (2023): 4987–5002, https://doi.org/10.1080/10447318.2023.2238367.

16. Heather Chen and Kathleen Magramo, "Finance Worker Pays Out $25 Million after Video Call with Deepfake 'Chief Financial Officer,'" CNN, February 4, 2024, https://www.cnn.com/2024/02/04/asia/deepfake-cfo-scam-hong-kong-intl-hnk/index.html. A "rug pull" is a type of financial scam in which developers of a cryptocurrency or NFT project suddenly withdraw all funds, abandon the project, and leave investors with worthless assets. See "What Is a Rug Pull and How to Avoid It?," Coinbase, accessed February 25, 2025, https://www.coinbase.com/en-nl/learn/tips-and-tutorials/what-is-a-rug-pull-and-how-to-avoid-it.

17. Sadia Idrees, Gianpaolo Vignali, and Simeon Gill, "Interactive Marketing with Virtual Commerce Tools: Purchasing Right Size and Fitted Garment in Fashion Metaverse," in *The Palgrave Handbook of Interactive Marketing*, ed. Cheng Lu Wang (Cham, Switzerland: Palgrave Macmillan, 2023), 329–351, https://doi.org/10.1007/978-3-031-14961-0_15.

18. Tanya Basu, "Cartier and Tiffany Are Getting into AR to Sell Luxury to Gen Z," *MIT Technology Review*, March 7, 2023, https://www.technologyreview.com/2023/03/07/1069414/cartier-tiffany-ar-luxury-gen-z/.

19. Arielle Pardes, "Try on Your Next Pair of Glasses Using Just Your iPhone," *Wired*, February 4, 2019, https://www.wired.com/story/warby-parker-augmented-reality-app/.

20. Thomas Kang, "Exploring the Potential of Virtual Commerce," *Walmart Corporate News*, September 19, 2023, https://corporate.walmart.com/news/2023/09/19/exploring-the-potential-of-virtual-commerce.

21. See "Ways to Shop with Google Assistant," Kroger, https://www.kroger.com/i/ways-to-shop/google-assistant; and Alexa Tietjen, "Exclusive: Snapchat Gives Kohl's Virtual Closet an Update," *WWD*, May 8, 2020, https://wwd.com/fashion-news/fashion-features/snapchat-kohls-virtual-closet-coronavirus-1203629594/.

22. "Revolutionizing Car Shopping: Virtual Showrooms in the Automotive Industry," Virtually Anywhere, May 29, 2024, https://virtually-anywhere.com/virtual-showrooms/revolutionizing-car-shopping-virtual-showrooms-in-the-automotive-industry/.

23. Do Yuon Kim, Ha Kyung Lee, and Kyunghwa Chung, "Avatar-Mediated Experience in the Metaverse: The Impact of Avatar Realism on User–Avatar Relationship," *Journal of Retailing and Consumer Services* 73 (2023): 103382, https://doi.org/10.1016/j.jretconser.2023.103382.

24. Kevan Yalowitz, Dwight Lee, Kevin Collins, Stephanie Gorski, Paul Johnson, and Brennan Torres, "Consumer Behavior in the Metaverse," Accenture, January 6, 2023, https://www.accenture.com/us-en/insights/software-platforms/metaverse-that-matters.

25. Kevan Yalowitz, Dwight Lee, Kevin Collins, Stephanie Gorski, Paul Johnson, and Brennan Torres, "Understanding Metaverse Consumer Behavior," Accenture, January 6, 2023, https://www.accenture.com/us-en/insights/software-platforms/metaverse-that-matters.

26. Yalowitz et al., "Consumer Behavior in the Metaverse."

27. Terry and Keeney, *The Metaverse Handbook*.

28. Fred Miao, Irina V. Kozlenkova, Haizhong Wang, Tao Xie, and Robert W. Palmatier, "An Emerging Theory of Avatar Marketing," American Marketing Association, March 9, 2021, https://www.ama.org/2021/03/09/an-emerging-theory-of-avatar-marketing/.

29. Christine Moy and Adit Gadgil, *Opportunities in the Metaverse* (JPMorgan Chase, 2022), https://www.jpmorgan.com/content/dam/jpm/treasury-services/documents/opportunities-in-the-metaverse.pdf.

30. "Banking in the Metaverse: The Next Frontier," *Accenture* (blog), February 23, 2022, https://bankingblog.accenture.com/banking-in-the-metaverse-the-next-frontier.

31. David Kadio-Morokro, "Finance in the Metaverse: Opportunities and a Roadmap—Part 1," Ernst & Young Consulting, March 6, 2023, https://www.ey.com/en_us/financial-services/finance-in-the-metaverse--opportunities-and-a-roadmap.

32. Naïma Aïdi, "Tourism and the Metaverse: Towards a Widespread Use of Virtual Travel?," *The Conversation*, August 16, 2022, http://theconversation.com/tourism-and-the-metaverse-towards-a-widespread-use-of-virtual-travel-188858; Fannelie Gerard, "Why the Metaverse (Really) Matters for Travel," *Accenture* (blog), September 15, 2022, https://web.archive.org/web/20221006013217/https://www.accenture.com/us-en/blogs/compass-travel-blog/metaverse-travel; Katrina Woznicki, "A New Kind of Tourism: Virtual Travel in the Metaverse," Verizon, May 12, 2022, https://www.verizon.com/about/news/virtual-travel-metaverse.

33. Maksim Godovykh, Carissa Baker, and Alan Fyall, "VR in Tourism: A New Call for Virtual Tourism Experience Amid and After the COVID-19 Pandemic," *Tourism and Hospitality* 3, no. 1 (2022): 265–275, https://doi.org/10.3390/tourhosp3010018.

34. Wai Han Lo and Ka Lun Benjamin Cheng, "Does Virtual Reality Attract Visitors? The Mediating Effect of Presence on Consumer Response in Virtual Reality Tourism Advertising," *Information Technology & Tourism* 22, no. 4 (2020): 537–562, https://doi.org/10.1007/s40558-020-00190-2.

35. Melanie Lieberman, "17 Facts About the Great Wall of China You Should Know," *Travel + Leisure*, March 27, 2023, https://www.travelandleisure.com/travel-tips/the-great-wall-of-china.

36. Chourouk Ouerghemmi, Myriam Ertz, Néji Bouslama, and Urvashi Tandon, "The Impact of Virtual Reality (VR) Tour Experience on Tourists' Intention to Visit," *Information* 14, no. 10 (2023): 546, https://doi.org/10.3390/info14100546.

37. Mark Twain, *The Innocents Abroad, or The New Pilgrim's Progress* (Hartford, CT: American Publishing Company, 1869), 650.

38. "Accenture Technology Vision 2022: 'Metaverse Continuum' Redefining How the World Works, Operates and Interacts," *Accenture Newsroom*, March 16, 2022, https://newsroom.accenture.com/news/2022/accenture-technology-vision-2022-metaverse-continuum-redefining-how-the-world-works-operates-and-interacts.

39. Ching-Hsuan Yeh, Yi-Shun Wang, Hsien-Ta Li, and Shuo-Yan Lin, "The Effect of Information Presentation Modes on Tourists' Responses in Internet Marketing: The Moderating Role of Emotions," *Journal of Travel & Tourism Marketing* 34, no. 8 (2017): 1018–1032, https://doi.org/10.1080/10548408.2016.1276509.

40. Kalin Bracken, "What Is Real Estate in the Metaverse? An Expert Explains," World Economic Forum, November 8, 2022, https://www.weforum.org/agenda/2022/11/property-buying-in-the-metaverse-is-the-future/; Raisa Bruner, "Why Investors Are Paying Real Money for Virtual Land," *TIME*, January 20, 2022, https://time.com/6140467/metaverse-real-estate/.

41. Matthew Sparkes, "What Is a Metaverse," *New Scientist* 251, no. 3348 (2021): https://doi.org/10.1016/S0262-4079(21)01450-0.

42. Michael Corridore, "The Metaverse: Revolutionizing the Way Government Agencies Function," Deloitte Canada, 2022, https://www2.deloitte.com/ca/en/pages/technology-media-and-telecommunications/articles/metaverse-considerations-government-public-service.html.

43. Riaz Pirmohamed et al., *The Metaverse: When Is Real Estate No Longer Real?* (London: Clifford Chance, October 2022), https://www.cliffordchance.com/content/dam/cliffordchance/briefings/2022/09/the-metaverse-when-is-real-estate-no-longer-real.pdf; see also Andrew Baker, "A Real Estate Boom in the Metaverse: Does It

Matter?," Ropes & Gray, March 29, 2022, https://www.ropesgray.com/en/insights/viewpoints/102hllw/a-real-estate-boom-in-the-metaverse-does-it-matter.

44. Debra Kamin, "The Next Hot Housing Market Is Out of This World. It's in the Metaverse," *New York Times*, February 19, 2023, https://www.nytimes.com/2023/02/19/realestate/metaverse-vr-housing-market.html; Debra Kamin, "Investors Snap Up Metaverse Real Estate in a Virtual Land Boom," *New York Times*, November 30, 2021, https://www.nytimes.com/2021/11/30/business/metaverse-real-estate.html.

45. Elizabeth Howcroft, "Virtual Real Estate Plot Sells for Record $2.4 Million," *Reuters*, November 24, 2021, https://www.reuters.com/markets/currencies/virtual-real-estate-plot-sells-record-24-million-2021-11-23/.

46. Josh Howarth, "75+ Metaverse Statistics (New 2024 Data)," *Exploding Topics* (blog), November 22, 2023, https://explodingtopics.com/blog/metaverse-stats.

47. Rohan Jambhale, "Metaverse Statistics by Market Capitalization, Revenue, Companies, Users, Meta Platform Income and Facts," Electro IQ, updated November 18, 2024, https://electroiq.com/stats/metaverse-statistics.

48. Megan Garber, "We've Lost the Plot," *The Atlantic*, January 30, 2023, https://www.theatlantic.com/magazine/archive/2023/03/tv-politics-entertainment-metaverse/672773/.

49. Lisa Respers France, "Travis Scott's Virtual Concert on *Fortnite* Set a Record," CNN, April 24, 2020, https://edition.cnn.com/2020/04/24/entertainment/travis-scott-fortnite-concert/index.html.

50. *Interoperability in the Metaverse* (World Economic Forum, January 2023), https://www3.weforum.org/docs/WEF_Interoperability_in_the_Metaverse.pdf.

51. Verena Fulde, "Our Vision for an Ethically Responsible Metaverse," Deutsche Telekom, December 22, 2023, https://www.telekom.com/en/company/details/our-vision-for-an-ethically-responsible-metaverse-1056850.

52. Bernard Marr, "The World of Metaverse Entertainment: Concerts, Theme Parks, and Movies," *Forbes*, July 28, 2022, https://www.forbes.com/sites/bernardmarr/2022/07/27/the-world-of-metaverse-entertainment-concerts-theme-parks-and-movies/.

53. Isamu Nishijima, "Ariana Grande x *Fortnite* Rift Tour: The Apogee of Pop Culture or Just the Beginning?," *Headline Asia*, August 28, 2021, https://headline.com/asia/en-us/post/ariana-grande-x-fortnite-rift-tour-the-apogee-of-pop.

Chapter 10

1. Tom Loozen and Adrian Baschnonga, "Seven Ways Telecom Operators Can Power the Metaverse," Ernst & Young Consulting, May 10, 2022, https://www.ey.com

/en_gl/insights/telecommunications/seven-ways-telecom-operators-can-power-the-metaverse.

2. Bronwyn Howell, "Governing the Metaverse: Can We Learn from Telecommunications?," *AEIdeas*, April 20, 2022, https://www.aei.org/technology-and-innovation/governing-the-metaverse-can-we-learn-from-telecommunications/.

3. Sydney Price and David DiMolfetta, "Big 4 Wireless Carriers Spent $100B on 5G Spectrum: Was It Worth It?," S&P Global, January 26, 2022, https://www.spglobal.com/market-intelligence/en/news-insights/articles/2022/1/big-4-wireless-carriers-spent-100b-on-5g-spectrum-was-it-worth-it-68488095.

4. "Understanding Hybrid 5G: Key Players and Trends," *Telecom Review Europe*, October 22, 2024, https://www.telecomrevieweurope.com/articles/reports-and-coverage/understanding-hybrid-5g-key-players-and-trends/.

5. DOCOMO, "DOCOMO and NTT Expand 6G Collaborations with SK Telecom and Rohde & Schwarz," press release, February 22, 2024, https://www.docomo.ne.jp/english/info/media_center/pr/2024/0222_00.html.

6. MTN Group, "MTN Group Extends the Magic of 5G Technology to Benin and Congo-Brazzaville," press release, November 25, 2024, https://www.mtn.com/mtn-group-extends-the-magic-of-5g-technology-to-benin-and-congo-brazzaville/.

7. "Vodafone and Meta's Collaboration Is a Glimpse into the Future of Mobile Streaming," Telecoms, September 9, 2024, https://www.telecoms.com/streaming-svod/vodafone-and-meta-s-collaboration-is-a-glimpse-into-the-future-of-mobile-streaming.

8. Jess Weatherbed, "Meta Is Building the 'Mother of All' Subsea Cables," *The Verge*, November 29, 2024, https://www.theverge.com/2024/11/29/24308746/meta-10-billion-global-subsea-cable-project; see also Ingrid Lunden, "Meta Plans to Build a $10B Subsea Cable Spanning the World, Sources Say," *TechCrunch*, November 29, 2024, https://techcrunch.com/2024/11/29/meta-plans-to-build-a-10b-subsea-cable-spanning-the-world-sources-say/; Gaya Nagarajan and Alex-Handrah Aimé, "Unlocking Global AI Potential with Next-Generation Subsea Infrastructure," *Engineering at Meta*, February 14, 2025, https://engineering.fb.com/2025/02/14/connectivity/project-waterworth-ai-subsea-infrastructure/.

9. Kelly Hill, "5G by the Numbers: 72 Countries, Nearly 2,000 Cities, 24 Standalone Networks," *RCR Wireless News*, May 3, 2022, https://www.rcrwireless.com/20220503/5g/5g-by-the-numbers-72-countries-nearly-2000-cities-24-standalone-networks.

10. Adaora Udoji, "The Metaverse Is Coming—It Just Needed 5G," Verizon, November 18, 2020, https://www.verizon.com/about/news/5g-makes-metaverse-real.

11. Ani Petrosyan, "Internet Usage Worldwide—Statistics & Facts," Statista, January 13, 2025, https://www.statista.com/topics/1145/internet-usage-worldwide/#topicOver

view; "Share of Time Spent Using the Internet on Mobile Phones per Day for Users Worldwide from 3rd Quarter 2013 to 2nd Quarter 2024," Statista, October 2024, https://www.statista.com/statistics/1289723/share-time-spent-mobile-internet-daily/; "Distribution of Average Daily Time Spent Online Worldwide from 3rd Quarter 2013 to 3rd Quarter 2024, by Device," Statista, February 2025, https://www.statista.com/statistics/1380539/time-spent-online-daily-by-device/.

12. *The Mobile Economy 2024* (London: GSMA, 2024), https://www.gsma.com/solutions-and-impact/connectivity-for-good/mobile-economy/wp-content/uploads/2024/02/260224-The-Mobile-Economy-2024.pdf; see also *The Mobile Economy China 2023* (London: GSMA, 2023), https://data.gsmaintelligence.com/api-web/v2/research-file-download?id=74384142&file=270323-Mobile-Economy-China-2023.pdf.

13. Antonino Masaracchia, Dang Van Huynh, George C. Alexandropoulos, Berk Canberk, Octavia A. Dobre, and Trung Q. Duong, "Toward the Metaverse Realization in 6G: Orchestration of RIS-Enabled Smart Wireless Environments via Digital Twins," *IEEE Internet of Things Magazine* 7, no. 2 (2024): 22–28, https://doi.org/10.1109/IOTM.001.2300128; Ruizhi Cheng, Nan Wu, Songqing Chen, and Bo Han, "Will Metaverse Be NextG Internet? Vision, Hype, and Reality," *IEEE Network* 36, no. 5 (2022): 197–204, https://doi.org/10.48550/arXiv.2201.12894.

14. Zhisheng Chen, "Metaverse Office: Exploring Future Teleworking Model," *Kybernetes* 53, no. 6 (2023): 2029–2045, https://doi.org/10.1108/K-10-2022-1432.

15. "The Global Gender Gap in Innovation and Creativity," World Intellectual Property Organization, 2024, https://www.wipo.int/about-ip/en/ip_innovation_economics/gender_innovation_gap/gender-parity-patenting.html; *Global Gender Gap Report 2022* (World Economic Forum, July 2022), https://www3.weforum.org/docs/WEF_GGGR_2022.pdf; Kenny Dooley, "Diversity & Inclusion in the Energy Sector," OGV Energy, July 6, 2021, https://www.ogv.energy/news-item/diversity-inclusion-in-the-energy-sector.

16. Based on data from the International Energy Agency (IEA). See official site for IEA stats, as well as stats for Organisation for Economic Co-operation and Development (OECD) and the world: "Gender and Energy Data Explorer," IEA 50, last updated November 26, 2024, https://www.iea.org/data-and-statistics/data-tools/gender-and-energy-data-explorer?Topic=Senior+Management&Indicator=Share+of+female+senior+managers+%28country+of+headquarters%29&Year=2023.

17. Chenghui Zhang and Shuai Liu, "Meta-Energy: When Integrated Energy Internet Meets Metaverse," *IEEE/CAA Journal of Automatica Sinica* 10, no. 3 (2023): 580–583, https://doi.org/10.1109/JAS.2023.123492; Christian Stoll, Ulrich Gallersdörfer, and Lena Klaaßen, "Climate Impacts of the Metaverse," *Joule* 6, no. 12 (2022): 2668–2673, https://doi.org/10.1016/j.joule.2022.10.013.

18. *Tracking Trends in a Time of Change: The Need for Radical Action Towards Sustainable Transport Decarbonisation, Transport and Climate Change Global Status Report*, 2nd

ed. (SLOCAT, 2021), https://unfccc.int/sites/default/files/resource/202202251552-‑‑SLOCAT%20Transport%20and%20Climate%20Change%20Global%20Status%20Report_Global%20Overview.pdf; "Transportation Emissions Worldwide—Statistics & Facts," Statista, January 23, 2025, https://www.statista.com/topics/7476/transportation-emissions-worldwide/.

19. Lizzy Rosenberg, "Even Though It's Virtual, the Metaverse Does Actually Impact the Environment," World Economic Forum, February 16, 2022, https://www.weforum.org/agenda/2022/02/how-metaverse-actually-impacts-the-environment/.

20. Zakaria Abou El Houda and Bouziane Brik, "Next-Power: Next-Generation Framework for Secure and Sustainable Energy Trading in the Metaverse," *Ad Hoc Networks* 149 (2023): https://doi.org/10.1016/j.adhoc.2023.103243.

21. April Miller, "What Does the Metaverse Mean for the Future of Energy Consumption?," *Earth.org*, March 10, 2022, https://earth.org/metaverse-energy-consumption/.

22. Karen Hao, "Training a Single AI Model Can Emit as Much Carbon as Five Cars in Their Lifetimes," *MIT Technology Review*, June 6, 2019, https://www.technologyreview.com/2019/06/06/239031/training-a-single-ai-model-can-emit-as-much-carbon-as-five-cars-in-their-lifetimes/.

23. Chaim Gartenberg, "Intel Thinks the Metaverse Will Need a Thousand-Fold Increase in Computing Capability," *The Verge*, December 15, 2021, https://www.theverge.com/2021/12/15/22836401/intel-metaverse-computing-capability-cpu-gpu-algorithms.

24. James Kobielus, "Is the Carbon Footprint of AI Too Big?," *InfoWorld*, August 6, 2020, https://www.infoworld.com/article/2259356/is-the-carbon-footprint-of-ai-too-big.html.

25. "Meta Is Committed to Reaching Net Zero Emissions Across Its Value Chain in 2030," US Chamber of Commerce, August 17, 2020, https://www.uschamber.com/climate-change/facebook-reduced-its-operational-greenhouse-gas-emissions-footprint-59-2019; CEE Multi-Country News Center, "Microsoft Is Committed to Achieving Zero Carbon Emissions and Waste by 2030," Microsoft, May 18, 2023, https://news.microsoft.com/en-cee/2023/05/18/microsoft-is-committed-to-achieving-zero-carbon-emissions-and-waste-by-2030/; Apple, "Apple Commits to Be 100 Percent Carbon Neutral for Its Supply Chain and Products by 2030," press release, July 21, 2020, https://www.apple.com/newsroom/2020/07/apple-commits-to-be-100-percent-carbon-neutral-for-its-supply-chain-and-products-by-2030/.

26. Margaret Osborne, "Bitcoin Could Rival Beef or Crude Oil in Environmental Impact," *Smithsonian Magazine*, October 3, 2022, https://www.smithsonianmag.com/smart-news/bitcoin-could-rival-beef-or-crude-oil-in-environmental-impact/.

27. "Bitcoin Energy Consumption Index," Digiconomist, 2024, https://digiconomist.net/bitcoin-energy-consumption.

28. *2015 Women in Manufacturing Study: Exploring the Gender Gap* (Deloitte Development, 2015), https://www2.deloitte.com/content/dam/Deloitte/us/Documents/manufacturing/us-mfg-women-in-manufacturing-2015-study.pdf.

29. Lynda Laughlin and Cheridan Christnacht, "Women in Manufacturing," US Census Bureau, October 3, 2017, https://www.census.gov/newsroom/blogs/random-samplings/2017/10/women-manufacturing.html.

30. John Coykendall, Kate Hardin, John Morehouse, and David R. Brousell, "Exploring the Industrial Metaverse," Deloitte Research Center for Energy & Industrials, September 13, 2023, https://www2.deloitte.com/us/en/insights/industry/manufacturing/industrial-metaverse-applications-smart-factory.html.

31. Alex Koohang et al., "Shaping the Metaverse into Reality: A Holistic Multidisciplinary Understanding of Opportunities, Challenges, and Avenues for Future Investigation," *Journal of Computer Information Systems* 63, no. 3 (2023): 735–765, https://doi.org/10.1080/08874417.2023.2165197.

32. *2022 IEEE Smartworld, Ubiquitous Intelligence & Computing, Scalable Computing & Communications, Digital Twin, Privacy Computing, Metaverse, Autonomous & Trusted Vehicles (SmartWorld/UIC/ScalCom/DigitalTwin/PriComp/Meta)* (Haikou, China: IEEE, 2022), https://doi.org/10.1109/SmartWorld/UIC/ScalC56740.2022.

33. Binil Starly, Pavel Koprov, Akshay Bharadwaj, Thomas Batchelder, and Bennett Breitenbach, "'Unreal' Factories: Next Generation of Digital Twins of Machines and Factories in the Industrial Metaverse," *Manufacturing Letters* 37 (2023): 50–52, https://doi.org/10.1016/j.mfglet.2023.07.021.

34. *The Effectiveness of Virtual Reality Soft Skills Training in the Enterprise* (PwC, June 25, 2020), https://www.pwc.com/us/en/services/consulting/technology/emerging-technology/assets/pwc-understanding-the-effectiveness-of-soft-skills-training-in-the-enterprise-a-study.pdf.

35. Luis Omar Alpala, Darío J. Quiroga-Parra, Juan Carlos Torres, and Diego H. Peluffo-Ordóñez, "Smart Factory Using Virtual Reality and Online Multi-User: Towards a Metaverse for Experimental Frameworks," *Applied Sciences* 12, no. 12 (2022): 6258, https://doi.org/10.3390/app12126258.

36. Gülçin Büyüközkan, "Metaverse and Supply Chain Management Applications," in *Metaverse: Technologies, Opportunities and Threats*, Studies in Big Data, ed. Fatih Sinan Esen, Hasan Tinmaz, and Madhusudan Singh (Singapore: Springer Nature Singapore, 2023), 383–395, https://doi.org/10.1007/978-981-99-4641-9_26; Alexandre Dolgui and Dmitry Ivanov, "Metaverse Supply Chain and Operations Management," *International Journal of Production Research* 61, no. 23 (2023): 8179–8191, https://doi.org/10.1080/00207543.2023.2240900.

37. See Siemensstadt, "Siemens Innovation Hub," https://www.siemensstadt.siemens.com/en; Jack Boreham, "Siemens Pioneers the Industrial Metaverse with Digital

Twin Technology: A Case Study," *Digital Twin Insider*, June 24, 2024, https://digital twininsider.com/2024/06/24/siemens-pioneers-the-industrial-metaverse-with-digital-twin-technology-a-case-study/; and Ling Fu, Jing Li, and Alex Greenberg, "Redefining Collaboration with the Industrial Metaverse and the Tecnomatix Portfolio," *Siemens Tecnomatix* (blog), July 29, 2024, https://blogs.sw.siemens.com/tecnomatix/redefining-collaboration-with-the-industrial-metaverse-and-the-tecnomatix-portfolio/.

38. Devin Liddell, "How the Metaverse Will Change Transportation as We Know It," *Fast Company*, November 9, 2021, https://www.fastcompany.com/90694275/how-the-metaverse-will-change-transportation-as-we-know-it.

Chapter 11

1. Yogesh K. Dwivedi et al., "Metaverse Beyond the Hype: Multidisciplinary Perspectives on Emerging Challenges, Opportunities, and Agenda for Research, Practice and Policy," *International Journal of Information Management* 66 (2022): 102542, https://doi.org/10.1016/j.ijinfomgt.2022.102542.

2. Mark Jamison and Matthew Glavish, "The Dark Side of the Metaverse, Part I," *AEIdeas*, March 17, 2022, https://www.Aei.Org/Technology-and-Innovation/the-Dark-Side-of-the-Metaverse-Part-i.

3. See Kim Parker, Juliana Menasce Horowitz, and Rachel Minkin, "COVID-19 Pandemic Continues to Reshape Work in America," Pew Research Center, February 16, 2022, https://www.pewresearch.org/social-trends/2022/02/16/covid-19-pandemic-continues-to-reshape-work-in-america/.

4. Rachel Pelta, "FlexJobs Survey Finds Employees Want Remote Work Post-Pandemic," *FlexJobs* (blog), April 19, 2021, https://www.flexjobs.com/blog/post/flexjobs-survey-finds-employees-want-remote-work-post-pandemic/; André Dua et al., "Is Remote Work Effective: We Finally Have the Data," McKinsey, June 23, 2022, https://www.mckinsey.com/industries/real-estate/our-insights/americans-are-embracing-flexible-work-and-they-want-more-of-it.

5. Jeanne C. Meister, "How Companies Are Using VR to Develop Employees' Soft Skills," *Harvard Business Review*, January 11, 2021, https://hbr.org/2021/01/how-companies-are-using-vr-to-develop-employees-soft-skills.

6. Emma Goldberg, "Here's What We Do and Don't Know About the Effects of Remote Work," *New York Times*, October 10, 2023, https://www.nytimes.com/2023/10/10/business/remote-work-effects.html.

7. Jeffrey W. Brown, *Leading the Digital Workforce: IT Leadership Peak Performance and Agility*, Security, Audit and Leadership Series (Boca Raton, FL: CRC Press, 2023); Aman Kumar, Amit Shankar, Abhishek Behl, Brij B. Gupta, and Sudha Mavuri, "Lights, Camera, Metaverse! Eliciting Intention to Use Industrial Metaverse,

Organizational Agility, and Firm Performance," *Journal of Global Information Management* 31, no. 8 (2023): 1–20, https://doi.org/10.4018/JGIM.333169.

8. Paola Barra et al., "MetaCUX: Social Interaction and Collaboration in the Metaverse," in *Human–Computer Interaction—INTERACT 2023*, Lecture Notes in Computer Science, v. 14145 (Cham, Switzerland: Springer Nature Switzerland, 2023), 528–532, https://doi.org/10.1007/978-3-031-42293-5_67; Patrick Hendriks, Christian M. Olt, Timo Sturm, and Clara C. Moos, "Exploring Team Collaboration in the Metaverse from a Human Capital Perspective," *Journal of Intellectual Capital* 25, no. 4 (2024): 686–710, https://doi.org/10.1108/JIC-02-2024-0055; Shu Schiller, Fiona Fui-Hoon Nah, Andy Luse, and Keng Siau, "Men Are from Mars and Women Are from Venus: Dyadic Collaboration in the Metaverse," *Internet Research* 34, no. 1 (2024): 149–173, https://doi.org/10.1108/INTR-08-2022-0690.

9. Tereza Šímová, Kristýna Zychová, and Martina Fejfarová, "Metaverse in the Virtual Workplace," *Vision: The Journal of Business Perspective* 28, no. 1 (2024): 19–34, https://doi.org/10.1177/09722629231168690; Jamadi, Bulan Prabawani, Widiartanto Widiartanto, and Reni Shinta Dewi, "Identification of Communication Trends in Business Practices Efforts to Increase Productivity and Smooth Communication in the Workplace: Review What Evidences Say," *Enrichment: Journal of Management* 12, no. 4 (2022): 2467–2478, https://doi.org/10.35335/enrichment.v12i4.677.

10. "Bringing the Benefits of In-Person Collaboration to the Virtual World," MIT, July 10, 2020, https://betterworld.mit.edu/bringing-the-benefits-of-in-person-collaboration-to-the-virtual-world/; and see "Meta Horizon Workrooms," Meta, https://forwork.meta.com/nl/en/horizon-workrooms/.

11. Ryan Golden, "BofA Banks on Virtual Reality to Train Workers for the 'Moments of Truth,'" *HR Dive*, March 16, 2021, https://www.hrdive.com/news/bofa-banks-on-virtual-reality-to-train-workers-for-the-moments-of-truth/596780/; Katherine Doherty, "Bank of America Is Using AI and Metaverse to Train New Hires," *Bloomberg*, July 13, 2023, https://www.bloomberg.com/news/articles/2023-07-13/bank-of-america-is-using-the-metaverse-ai-to-train-its-hires.

12. Omaima Hajjami and Sunyoung Park, "Using the Metaverse in Training: Lessons from Real Cases," *European Journal of Training and Development* 48, no. 5/6 (2024): 555–575, https://doi.org/10.1108/EJTD-12-2022-0144.

13. Dilek Kitapcioglu, Mehmet Emin Aksoy, Arun Ekin Ekin, and Tuba Usseli, "Comparing Learning Outcomes of Machine Guided VR-Based Training with Educator Guided Training in Metaverse Environment: A Randomized Controlled Study (Preprint)," *JMIR Serious Games* (2024), https://doi.org/10.2196/58654; Tai-Cheng Liu, Ai-Shi Liu, Zhi-Gang Bai, and Lei Zhao, "The Metaverse Training Room for Cardiovascular Interventional Surgery," *Asian Journal of Surgery* 46, no. 7 (2023): 2780–2781, https://doi.org/10.1016/j.asjsur.2023.01.043.

14. Eman AbuKhousa, Mohamed Sami El-Tahawy, and Yacine Atif, "Envisioning Architecture of Metaverse Intensive Learning Experience (MiLEx): Career Readiness in the 21st Century and Collective Intelligence Development Scenario," *Future Internet* 15, no. 2 (2023): 53, https://doi.org/10.3390/fi15020053; Ahmed Tlili et al., "Is Metaverse in Education a Blessing or a Curse: A Combined Content and Bibliometric Analysis," *Smart Learning Environments* 9 (2022): 24–31, https://doi.org/10.1186/s40561-022-00205-x.

15. Bank of America, "Bank of America Is First in Industry to Launch Virtual Reality Training Program in Nearly 4,300 Financial Centers," press release, October 7, 2021, https://newsroom.bankofamerica.com/content/newsroom/press-releases/2021/10/bank-of-america-is-first-in-industry-to-launch-virtual-reality-t.html.

16. Mark Purdy, "How the Metaverse Could Change Work," *Harvard Business Review*, April 5, 2022, https://hbr.org/2022/04/how-the-metaverse-could-change-work.

17. Danielle Abril, "What You Need to Know About the Metaverse and the Future of Work," *Washington Post*, March 4, 2022, https://www.washingtonpost.com/technology/2022/03/04/metaverse-future-of-work-virtual-augmented-reality/.

18. Javier Gonzalez Nuñez and Manuel Bolognesi, "Exploring Team Collaboration in the New Metaverse (The 3D-AI Internet)," *SocioEconomic Challenges* 8, no. 2 (2024): 314–341, https://doi.org/10.61093/sec.8(2).314-341.2024.

19. Riichiro Mizoguchi, Pierre Dillenbourg, and Zhiting Zhu, eds., *Learning by Effective Utilization of Technologies: Facilitating Intercultural Understanding*, Frontiers in Artificial Intelligence and Applications, v. 151 (Amsterdam: IOS Press, 2006).

20. Jennifer Korn, "Hackers Steal over $600 Million from Video Game Axie Infinity's Ronin Network," CNN, March 29, 2022, https://www.cnn.com/2022/03/29/tech/axie-infinity-ronin-hack/index.html.

21. Stephanie Sierra, "Bay Area Investor Loses $1.2M in Crypto Scam as Fraud Cases Triple Across CA," ABC7 News, July 27, 2022, https://abc7news.com/crypto-scam-pig-butchering-fbi-silicon-valley/12077745/.

22. Adecco Group, "Work in the Metaverse? How the Metaverse Is Shaping the Future of Work," Adecco Group, February 9, 2022, https://www.adeccogroup.com/future-of-work/latest-insights/how-the-Metaverse-is-shaping-the-future-of-work/.

23. Shane L. Rogers, Rebecca Broadbent, Jemma Brown, Alan Fraser, and Craig P. Speelman, "Realistic Motion Avatars Are the Future for Social Interaction in Virtual Reality," *Frontiers in Virtual Reality* 2 (2022): 163, https://doi.org/10.3389/frvir.2021.750729.

24. See "Immutable Games Spotlight: IMVU Is the World's Biggest Web3 Social Metaverse," Immutable, July 21, 2023, https://www.immutable.com/blog/immutable-games-spotlight-imvu-is-the-worlds-biggest-web3-social-metaverse.

25. Phil Rosen, "This Metaverse Casino Raked in $7.5 Million in Last 3 Months and Accounts for a Third of Decentraland's Daily Users," *Markets Insider*, February 4, 2022, https://markets.businessinsider.com/news/currencies/decentraland-metaverse-casino-crypto-gambling-poker-2022-2.

Conclusion

1. Paola Cecchi-Dimeglio, Taha Masood, and Andy Ouderkirk, "What Makes Innovation Partnerships Succeed," *Harvard Business Review*, July 14, 2022, https://hbr.org/2022/07/what-makes-innovation-partnerships-succeed.

Index

Page numbers followed by *t* indicate tables.

Abed, Gabriel, 138
Accenture, 98, 160, 171
Activision Blizzard, 31, 247n7
Adidas, 14, 28, 208
AECOM, 173
Affirmative action, 125
AI (artificial intelligence). *See also* LLMs; NLP
   adoption across sectors, 6
   advancement of, 4
   algorithmic bias in, 48, 53
   data analytics' integration with, 219
   educational role of, 141–142
   empathetic, 70
   energy use by, 191
   generative, 4, 217–218
   importance to businesses, 6
   information moderation by, 57
   and machine learning, 4, 34, 53, 119, 147
   metaverse's dependence on, 4, 17, 19, 223–225
   myths about, 13
   and telecoms, 187
AI: Discovering the Citiverse, 7, 18
Airbus, 182
Alexa, 35
Algorithmic valuation, transparent, 44
Alibaba, 183
AltspaceVR, 126
Amazon, 47, 49, 81, 92–93, 109, 119, 125–126, 201
Amazon Sumerian, 168
American Express, 164
American Tower Corporation (ATC), 97
Amnesty International, 135
Amphia (Netherlands), 148
Android operating system, 121, 130
Anxiety, 53, 94, 109, 126, 147–148, 150–151
AOL, 184
APIs (application programming interfaces), 140, 278n14
Apollo Hospitals, 97
Apple, 3, 69, 81, 95–96, 109, 121, 125–126, 192
AR (augmented reality), 4, 17
   defined, 26
   failures and risk-taking in, 84, 86–87
   glasses for, 16, 127
   investment in, 155
   notable failures in, 90
   Pokémon GO, 4, 26, 87
   popularity of, 11
Archie, 12
AR Place, 162
Artificial intelligence. *See* AI
ATC (American Tower Corporation), 97

AT&T, 183
Augmented reality. *See* AR
Autism spectrum disorder (ASD), 151
Avatars
  and body image, 59–60
  business-to-avatar (B-2-A) marketing, 162–163
  corporate, 20
  customer feedback on, 126–127
  defensive maneuvers for, 76
  defined, 12, 20
  ethical issues surrounding, 21, 55
  mental health concerns addressed via, 150–151
  multilingual, 138
  privacy and anonymity via, 21, 74
  purpose of, 20
  safe zones for, 76
  self-representation and self-expression via, 20–21, 74
  types of, 20
  user base represented by, 33
  user-created, customized, 4, 11, 20, 55–56, 74
  users prepped for metaverse via, 37
Axie Infinity, 216
Azure, 22

Baby boomers, 28, 195–196
Bajaj Auto (India), 182
Banking/financial services, 153, 163–167
Bank of America, 208, 211–212
Barbados, 137–138
Bard, 4
Barella, Leo, 98
Behavioral economics, 60
Behavioral mapping, 82
Behavioral science, 6–7, 56, 59–60, 80–83, 128–130, 142, 193–195
Behavioral shifts, 51–63
  avatar customization, 55–56
  avatars and body image, 59–60

continued challenges, 52–53
customer well-being, 53–54, 56, 58
and digital identity, 55
extremism, 59
facing changes, 51–52
healing from Web 2.0, 54–56
intentional design, 1, 48, 60–61, 70
on the internet, 51
mental and physical health, 52–56, 59–60, 62
the metaverse as reshaping us, 51–52, 61–63
physical exercise, 53–54
reducing adverse behaviors, 56–57
skepticism, 61
sleep, 54
stress management, 54
technology addiction, 53–54, 223
trust, 56–59
worrying, 60
Betamax, 89
Bezos, Jeff, 89, 92
Bias
  algorithmic, 48, 53
  cognitive, 71–73, 77, 83
  cognitive, and decision-making, 71
  confirmation, 71–73, 77, 83
  echo chamber effect, 33, 71–72, 82, 131
  fundamental attribution error, 73
  gender, 52–53
  halo effect, 73
  horn effect, 73
  in-group, 74
  innovation impacted by, 69–74
  intersectional, 232, 237
  leaders' recognition and neutralization of, 76–77
  in LLMs, 231–239*tt*, 249n18
  from logical errors/flawed group thinking, 71–72
  in metaverse design, 34
  overcoming, 74

overconfidence, 71–72, 82
perception shaped by, 73–74
racial, 44, 126
structural, 69
systemic, 55, 71
unconscious, 69
Big data, 6, 131
Bitcoin, 25, 27, 188, 193
Black Women Online (BWO), 241n3
Blockbuster, 207
Blockchain. *See also* Cryptocurrencies; Ethereum; NFTs
and decentralized networks, 13, 15, 24–25, 27
defined, 24
evolution of, 13, 15
role in metaverse, 17, 24–25, 43
security of, 139, 151, 216
Bloomberg Intelligence, 31
Bloomfield, Robert, 250n27
Blu-ray, 89, 119
BMW, 107, 181–182, 196–197, 202
Boeing, 181–182, 196
Bosch, 220
BP, 98–99
Braille smartwatches, 79–80
BT Group, 183
Buckmire, Susan, 241n3
Built spaces, 174
Business, future of, 221–222
BWO (Black Women Online), 241n3

Call of Duty, 247n7
Carbon emissions, 171, 188–191, 193, 195, 223
Cardiothoracic Surgery Research Training Program (UPMC), 281n44
CarMax, 159
Chatbots, 4
ChatGPT, 4
Chief collaboration officers, 108
Chief metaverse officers, 157

China Mobile, 183, 186
Cities, digital twins of, 135–137
Classcraft, 145
Cleveland Clinic, 148–149
Cline, Ernest: *Ready Player One*, 13
Cloud computing services, 119–120
Clubhouse, 163
Coca-Cola, 47
Collaboration. *See also* Partnerships
and communication, 6, 207–208
among inventors, 79, 107–108, 113
and trust, 121–125
College of Nursing (University of Central Florida), 148
Colleges, metaverse's impact on, 145–146
Collins, Patricia Hill, 241n3
Commerce, 11, 31, 153–163
Committee on Accessibility and Inclusion (ITU), 7, 18
Communication
and collaboration, 6, 207–208
and innovation, 100–103, 105–107, 109–110, 112–117
in partnerships, 122–123
and trust, 121–122
Communications service providers (CSPs), 187
Compaq, 119
Confirmation bias, 71–73, 77, 83
ConsenSys, 252n48
Construction. *See* Real estate, construction, and planning
Consumer engagement
and behavioral science, 194
and innovation, 109, 126–127, 130
Cookies, 58, 158, 162
COVID-19 pandemic
lockdowns during, 11, 36
misinformation during, 57
online learning during, 143–144
online shopping during, 32
remote work during, 206

COVID-19 pandemic (cont.)
role in the metaverse's evolution, 11, 156, 165–166, 225
tourism impacted by, 170
work–life balance during, 212
Creativity. See Innovation
Cryptocurrencies. See also Blockchain
bitcoin, 25, 27, 188, 193
defined, 27
fraud in, 158, 284n16
MetaMask, 48, 252n48
metaverse wallet for, 27, 48
theft of, 216
for virtual casinos and sports betting, 42, 221
CryptoPunks, 43–44
CSIR (South Africa), 182
Curb ramps, 79
Customer feedback, 125–127, 131
Customer service, virtual, 138–139
Cyberbullying, 51, 109
Cybersecurity, 216

DaimlerChrysler, 122
DAQRI, 90
Data
big, 6, 131
biometric, 17, 48, 54, 56, 58, 139
breaches of, 139 (see also under Privacy and security)
EEO-1 Component 1 report, 258n9
eye-tracking, 48
misuse of, 58, 147
and network latency (see Latency)
Data analytics, 6, 19, 219
DEC (Digital Equipment Corporation), 119
Decentraland, 41, 47, 96, 124, 137–138, 164, 177, 217, 249n22, 252n46
Decentralized applications (dApps), 26, 252n48
Decentralized finance (DeFi), 221

Decentralized networks
and blockchain, 13, 15, 24–25, 27
purpose of, 27
Delta Air Lines, 210
Democritus, 19
Department of Defense, 140
Department of Energy, 188
Department of Transportation, 136, 174
Department of Veterans Affairs, 150–151
Depression, 53–54, 126, 150
Design, 95–117. See also Innovation
behavioral science applied to, 129
bias in, 34
collaborative, 197–198
core competencies and focus, 95–96, 224
ethical, 56
inclusive, 22, 32, 48, 52, 58, 61, 83, 103–104, 130–131, 224–225
intentional, 1, 48, 60–61, 70
multilayered and flexible strategy for, 97–99
participation gaps closed via, 104–105
precedents set by, 55
proactive oversight in, 53
of a purpose-driven campaign, 111
of realistic virtual prototypes, 197
strategic, 22, 61
success, defining, 96–97
user-focused, 3, 5, 70–71
Deutsche Telekom, 179
DHL, 210
Digital ecosystems, 5, 7, 104, 130
Digital Equipment Corporation (DEC), 119
Digital-first economies and strategies, 5, 8, 23
Digital ID, 75
Digital twinning, 38–39, 136–137, 140, 171–172, 174, 178, 187, 189–190, 198–202
Digital whiteboards, 219

Diplomacy, 137–138
Disney, 124, 157, 170
Disney, Walt, 89
Diversity. *See* Inclusivity
Diversity Pledge, 84
Dolce & Gabbana, 3
Dove, 3
DressX, 163
Drones, 201
Dubai, 136

EBay, 47
Echo chamber effect, 33, 71–72, 82, 131
E-commerce, 11, 31, 153–163
Ecosystems, 35–36
Edmondson, Amy, 87
Education, 141–147, 210–211. *See also* Immersive learning/training
Education 4.0 Alliance (Reskilling Revolution), 141–142
EEOC (Equal Employment Opportunity Commission), 68, 258n9
Electric cars, 89
Electricity consumption, 193
Electronics and Telecommunications Research Institute (ETRI; South Korea), 137
Embraer (Brazil), 182
Emojis, 35
Empathy, 55, 74, 175
Energy
  companies, 188–190
  consumer adoption of energy-saving practices, 194
  consumption of, 181, 188–193, 195
Entertainment and media, 153, 172–179
Environmental issues. *See* Carbon emissions; Energy
EON Reality, 145
Epic Games, 120, 124, 156
Equal Employment Opportunity Commission (EEOC), 68, 258n9
Ericsson, 122–123

EssilorLuxottica, 131
Ethereum, 25–27, 37, 41, 252n46, 252n48
Ethereum Request for Comments 20 (ERC-20), 26
Ethernet, 119
Ethics, 21, 55–56, 128, 146–147, 215, 219, 221–222
European Union (EU), 67
Extended reality (XR), 24, 26–27
ExxonMobil, 210

Facebook (*later* Meta). *See* Meta
Facial recognition software, 80
Factories, 198–199
Failure
  complex, 88
  fear of, 86
  intelligent, 88
  notable failures, 89–90
  not failing as not trying, 86–87
  as an option, 84–86
  safe-to-fail innovative culture, 84–85, 87–89, 93, 131
  simple, 87–88
  supporting, 88–89
Fashion, 33, 37, 41
Fast and slow thinking, 91
FedEx, 120, 201
FEMA (Federal Emergency Management Agency), 174
FG-MV (Focus Group on the Metaverse; ITU), 7
Financial services/banking, 153, 163–167
Finding Nemo (film), 91
Fintech (financial technology), 33
First responders, training of, 139
Fisher, Donna, 241n3
Fisk University, 145
Fitbit, 92
5G networks, 17, 183–184, 186
FlickPlay, 136

Flip phones, 12
Floreo, 104–105, 150–151
Florida A&M University, 145
Focus Group on the Metaverse (FG-MV; ITU), 7
Foo Fighters, 168
Ford, 47, 181–182, 212, 220
Fortnite, 48, 157, 169, 178–179
Friends with Holograms, 74

Gaming
  consumer demographics of, 75–76
  engagement for, 35–36
  growth of, 42
  platforms' investment in metaverse, 37
  role in starting the metaverse, 13–14
Gap, 14
Gartner, 31
GDPR (General Data Protection Regulation), 236–238
GE (General Electric), 47, 212
GE Aerospace, 182, 196–197
GE HealthCare, 182, 196–197
Gender diversity, 188, 196, 289n16
General Data Protection Regulation (GDPR), 236–238
General Electric (GE), 47, 212
Genius Bar, 109
Gen X, 28
Gen Z, 16, 28, 152, 156
Gestamp, 185
Gibson, William: *Neuromancer*, 13
Global Initiative on Virtual Worlds (ITU), 7, 18
Global Mercy digital twin, 99
Gmail, 91
Goldman Sachs, 31
Google
  Apple's collaboration with, 121
  behavioral science used by, 81
  customer feedback used by, 125
  early adoption of metaverse by, 47
  growth of, 49
  Meta's collaboration with, 119–120
  psychological safety at, 91
  role in building the metaverse, 28
  Unity's collaboration with, 124
  workforce representation at, 69
Google Assistant, 35
Google Cardboard, 90
Google Cloud, 120
Google Daydream, 90
Google Expeditions, 143
Google Glass, 87, 128–129, 131
Google Search, 91
Google Street View, 200
Google Vision Pro, 17
Google Workspace, 119
Governance, 135–136
  creating structures of, 23
  of future work, 215
  ITU's role in, 7, 18
  of the metaverse economy, 42, 44
Governments, 136–141, 278n14
GPS, 119
Grande, Ariana, 168, 179
Greenhouse gas emissions, 191
GSMA Intelligence, 186
Gucci, 157
Gucci Garden, 157

Hackers, 216
Harlem (New York City), 142
Harmony Hub, 76
Harrison, Suzanne, 84
Harvard Kennedy School, 228–229
Harvard Law School, 228
Harvard University, 6
HD DVD, 89
Healing, distance, 151–152
Health care, 97, 135, 148–152, 160, 281n44
Hearing aids, smart, 79
Heilig, Morton, 46
Hewlett Packard Enterprise (HPE), 120

Highly Immersive Visualization
  Environments (HIVEs), 98–99
HIPAA, 56
Hirameki Garden, 98
HIVEs (Highly Immersive Visualization
  Environments), 98–99
H&M, 3, 159
HOK, 173
Holograktor, 202
Homelink, 163–164
Honeywell, 212
Hong Kong, 136
Horizon, 163, 209
Horizon Workrooms, 197–198
Horizon Worlds, 19, 76, 126–127
HSBC, 164
HTC Vive, 14, 90
HTML, 45–46
Huang, Jensen, 86, 141–142
Hulu, 14

IBM, 47, 119–120
ICE Poker, 221
ID (identification), legal, 75
IDEO, 91–92
I-Glasses, 90
IKEA, 162
Immersive learning/training, 142–149,
  199–200, 209–210
Immersive vehicle operation, 200,
  202
IMVU, 217
Inclusivity, 223. *See also under*
  Innovation
  and accessibility, 34
  creators, diversity of, 34–35
  funding by men-led vs. women-led
    companies, 33
  inclusive design, 22, 32, 48, 52, 58,
    61, 83, 103, 130–131, 224–225
  women's representation in digital
    spaces, 33–34
Industrial Revolution, 223

Industry 4.0 (Fourth Industrial
  Revolution), 19
Innovation, 79–94. *See also* Design;
  Failure
  accessible, 79–80, 83–84, 130
  and anonymous idea submission,
    106–107
  and behavioral science, 80–83
  biases' impact on, 69–74
  collaborative, 79, 107–108, 113
  and communication, 100–103, 105–
    107, 109–110, 112–117
  and confirmation bias, 83
  and customer feedback, 125–127, 131
  diverse voices among inventors, 113,
    116–117
  and the echo chamber effect, 33, 72,
    82, 131
  elements of, 79
  engagement barriers, 112–113
  engagement by employees, 95, 100–
    105, 107, 109–111, 225
  engagement by users, 109, 126–127,
    130
  examples of, 79–80
  hackathons, 79, 85
  importance of, 79, 95
  inclusive, 67, 76–77, 79–80, 99, 125–
    126, 262n4
  information hubs for, 105–106
  and intuitive thinking, 90–91
  and IP, 99–100, 105, 107, 111, 115
  leaders' role in, 80, 84, 92–94, 103–
    105, 110–111, 116–117, 225
  and messaging, 110–113
  minority inventors, 83–84, 111
  and patents, 99–102, 104, 106, 111,
    115–116
  peer-driven, 114
  peer educators for, 115–116
  peer mentoring for, 114–115
  pitfalls for, 82–83
  and the planning fallacy, 82–83

Innovation (cont.)
  and psychological safety, 91–92, 94, 106, 110
  reaching your innovators, 108–109
  representation gaps in, 83–84
  user-driven, 79
  women inventors, 83–84, 100, 111
Instagram, 35
Intel, 47, 119, 191–192
Intellectual property. *See* IP
Interior design, 176
International Energy Agency (IEA), 289n16
International Telecommunication Union (ITU; United Nations), 7, 18
Internet. *See also* Web 2.0; Web 3.0
  adverse behaviors on, 56
  behavioral shifts on, 51
  as changing us, 51
  economy of, 1, 241n3
  evolution/trajectory of, 1, 45, 52, 241n1
  inclusivity of, 52, 241n3
  marketing via, 154
  vs. metaverse, 48
  peak hours on, 193
  vs. Web 3.0, 52–53
Intersectional bias, 232, 237
Intuit, 125
Intuitive thinking, 90–91
IP (intellectual property), 99–100, 105, 107, 111, 115, 121–125, 141
Iphone, 109, 119, 126, 130
Isolation, feelings of, 146, 208, 213
ISPs, first commercial, 241n1
ITU (International Telecommunication Union; United Nations), 7, 18

Jetsons, The (TV show), 12
Jobs, Steve, 89
Johns Hopkins Hospital, 148
Johnson, Ron, 109
JPMorgan Chase, 96, 164

Kahneman, Daniel, 91
Kant, Immanuel, 244n29
KB Bank (South Korea), 164
Kohl's, 159
Korea. *See* Republic of Korea
Korea Information Society Development Institute, 137
Kroger, 159

Large language models. *See* LLMs
Latency, 35, 37, 169–170, 181, 183, 185, 191, 197–198
Law enforcement training, 139
Leaders, female vs. male, 22–23
LEGO, 156
Leibniz, Gottfried Wilhelm, 244n29
Leonardo DaVinci, 169
LGBTQ+ community, 241n3
Liminal VR, 150
Linden Lab, 21–22
LLMs (large language models), 34, 70, 231–239tt, 249n18, 262n4
Lockheed Martin, 181–182
London, 136
Luckey, Palmer, 87

Machine learning (ML), 4, 34, 53, 119, 147
Madrid, 169
Magic Leap One, 87, 90
Manufacturing industry, 195–203
Market-driven accessibility, 5, 21–22
Marketing, 153–158
Market leadership, 22–23
Marvel, 124
Massachusetts Institute of Technology (MIT), 45, 144
Massively multiplayer online role-playing games (MMORPGs), 27
Matrix movies, 13
Matterport, 168, 172
Maxwell, K. C., 108
Mayo Clinic, 148–150

McDonald's, 153
McKinsey & Company, 31–32, 68
MDGs (Millennium Development Goals), 67
Médecins Sans Frontières (Doctors Without Borders), 135
Media and entertainment, 153, 172–179
Medical education, 148–149, 281n44
Medivis, 220
Mental health, 150–151, 213–214
Mercy Ships, 99
Meta (*formerly* Facebook), 11, 16–17, 28, 31, 169
  avatars offered by, 37
  customer feedback used by, 126–127
  energy companies' collaboration with, 188
  EssilorLuxottica's collaboration with, 131
  Google's collaboration with, 119–120
  Horizon Worlds, 19, 76, 126–127
  marketing on, 157
  Oculus acquired by, 33, 129
  power consumption by, 191–192
  and telecoms, 183
  and Verizon, 185
  VR use by, 208
  workforce representation at, 69
Metajuku, 164
MetaMask, 27, 48, 252n48
Metanomics, 39–43, 45, 250n27
Meta Quest Pro, 119–120
Metaverse
  adaptability of, 1–2
  addiction to, 62
  adoption timeframe for, 28–29
  as AI-dependent, 4, 17, 19, 223–225
  Americans' interest in and perceptions of, 40–41
  anticipated growth and early adopters of, 7–8, 16, 32, 40, 45, 47–49
  awareness of, 40–41
  blockchain's role in, 17, 24–25, 43
  broad market reach of, 1–3
  clothing, housing, and other essentials replicated by, 41–42
  consumer demographics of, 40–41, 75–76, 157
  consumer expectations of, 36–38, 249n22
  consumers' experience over the years, 4–5
  consumers' time spent in, 15–16
  digital assets in, 25 (*see also* NFTs)
  economic sectors impacted by, 46–47
  economic viability of, 17
  economies of, 1, 17, 23, 42–45, 131, 216–217
  evolution/growth of, 1, 11–13, 18–19, 28, 38, 40, 45–46, 224–225
  in fiction, 13, 15
  gaming's role in starting, 13–14
  generative, 4, 6, 19–20, 142, 217–218
  gig work in, 41–42
  goal-setting for, 37
  Google searches for, 38
  governance of, 18
  governmental/institutional interest in, 18 (*see also* Governments)
  harassment and abuse in, 22, 76
  headsets for, 16, 20
  as here and still arriving, 38, 249n22
  vs. holodeck, 12, 52, 242n4
  human component of, 48
  initial phase for businesses, 45
  interconnections, recognizing, 225
  vs. internet, 48
  as interoperable, integrated, and interactive, 27
  investments in, 17, 31–33, 38, 46
  misinformation in, 57–58
  monetization of, 7, 16, 31, 33, 36, 38
  as more than novel tech, 15
  myths about, 13–17
  navigating, key strategies for, 224–225

Metaverse (cont.)
  as a network of virtual platforms, 19–20
  origin of, 11, 13, 225
  overview of, 223–224
  perceptions of, 12, 15–16
  potential of, 17–18, 36, 46–47
  profitability of, 1, 5, 38
  questions to ask about, 3
  vs. real life, 14
  regulation of, 43–45, 62–63, 141
  re-personalization of activities and processes, 5, 37, 42, 166–167
  retrofitting of platforms, 21–22
  risks of, 1
  scaling of, 38
  societal change via, 48
  standardization of, 18
  status symbols in, 41
  strategic design of platforms, 22
  sustainability of, 1–2
  tech companies as essential to, 17 (*see also* AR; MR; VR)
  tech for accessing, 16–17, 20, 28
  terminology of, 19–20, 23–24, 27–28
  testing and simulating products and processes, 37–39, 42 (*see also* Digital twinning)
  thriving future via, requirements for, 1–2
  tokens in, 24–27, 166 (*see also* NFTs)
  as transformative, 12, 181, 188
  users as stakeholders in, 23
  value and growth of, 31–32, 39 (*see also* Metanomics)
  vs. VR, 14
Metaverse Learning, 220
Metaverse Seoul, 137
Metaverse wallet, 27, 48
Microservice architectures, 140, 278n14
Microsoft
  Activision Blizzard acquired by, 31
  avatars offered by, 37
  construction and urban planning software by, 173
  core focus of, 96
  customer feedback used by, 126–127
  digital twins used by, 196–197
  Fail Fast program at, 92
  FedEx's collaboration with, 120
  growth mindset at, 92
  Hewlett Packard's collaboration with, 120
  HIVEs project, 98–99
  Kin phone, 131
  Nokia's collaboration with, 122–123
  power consumption by, 192
  role in building the metaverse, 28, 31
  Samsung's collaboration with, 119
  Sony's collaboration with, 124
  and telecoms, 183
  USB technology, role in, 119
  VR use by, 208
  workforce representation at, 69
Microsoft HoloLens, 87, 90, 220
Microsoft Mesh, 11, 17, 22, 98, 127, 163, 208
Microsoft Teams, 98, 120, 208
Military applications of the metaverse, 140
Millennials (Gen Y), 16, 28, 152, 156
Millennium Development Goals (MDGs), 67
Mindset
  forward-thinking, 156
  growth, 88–92
  inclusive, 93
  innovation, 106–110, 194
  strategic, 161
Minecraft, 4
Ministry of Foreign Affairs and Foreign Trade (Barbados), 137–138
Ministry of Science and ICT (South Korea), 136
Mirror stereoscopes, 46

Misinformation, 1, 29, 51, 57–58, 218, 223
MIT (Massachusetts Institute of Technology), 45, 144
Mixed reality. *See* MR
MMC (Netherlands), 148
MMORPGs (massively multiplayer online role-playing games), 27
Monetization, 7, 16, 31, 33, 36
Mosaic, 12
Mozilla Hubs, 145
MR (mixed reality), 4, 17, 22, 24, 26, 84–85, 130, 148. *See also* Microsoft Mesh
MTN Group, 183
MTR Corporation Limited, 136
MUD1, 13
Multiverse, 19
Musk, Elon, 93

Nadella, Satya, 92
NASA, 125
National Archives and Records Administration (NARA) data breach, 139
National Institutes of Health (NIH), 126, 281n44
Natural language processing. *See* NLP
Nearpod, 145
Nest Thermostat, 92
Netflix, 47, 49, 124, 126, 179
Netscape, 184
Neuromancer (Gibson), 13
Newton personal digital assistant (PDA), 89
New York City, 136, 142
NFTs (non-fungible tokens)
 authenticity of, 25–26, 43
 benefits of, 163
 buying, selling, or exchanging, 27, 33
 counterfeit, 158
 economic balance in, 43–44
 Ethereum platform for, 25
 gender and racial bias in, 44
 introduction of, 17, 165
 marketing via, 154
 minting of, 27
 smart contracts covering, 26
Nike, 3, 28, 126, 157
Nike+iPod, 124
Nikeland, 153
Nintendo, 85, 90
Nissan, 202
NLP (natural language processing), 70, 233t, 239t, 262n4
Nokia, 122–123
Non-fungible tokens. *See* NFTs
Nottingham Building Society (UK), 163–164
Nova AI, 4
NTT Docomo, 179, 183
Nvidia, 19, 120, 124, 196–197, 208

Oculus, 33, 129–130
Oculus Go, 90, 129
Oculus Quest, 84–85, 129–130, 192
Oculus Quest 3, 127, 129–130
Oculus Rift, 14, 87, 90, 129
Omniverse, 107, 127, 208
One Accenture Park, 98
Online addiction, 51, 58
Onyx Lounge, 164
Opportunity, unlocking, 67–77
 bias as shaping perception, 73–74
 biases, overcoming, 74
 biases from logical errors/flawed group thinking, 71–72
 biases' impact on innovation, 69–74
 decision-making and cognitive biases, 71
 engagement vs. safety, 75–76
 inclusive innovations for, 67, 76–77
 lack of legal identification as a challenge, 75
 leaders' role in, 76–77
 workforce growth, 68–69

Optus, 183
Orange, 183, 185
Overwatch, 247n7

Pain management therapy, 148
Panasonic, 119, 124
Panorama 360, 200
Parkinson's disease, 149
Partnerships, 119–131
   benefits and risks of, 119–121, 124–125
   clear goals in, 121–122
   communication in, 122–123
   dispute resolution for, 124
   examples of, 119–123
   Google Glass (case study), 128–129
   money's role in, 123
   as the norm, 120–121
   Oculus (case study), 129–130
   responsibilities in, 122
   and shared accountability, 195
   with telecoms, 186 (*see also* Telecoms)
   trust in, 121–123
Patents, 99–102, 104, 106, 111, 115–116
PatientsLikeMe, 150
People Culture Data Consulting Group, 6
Pepsi, 3
Pew Research Center, 126
Philips, 119
Pinterest, 35
Pixar, 91, 124
PlayStation, 120
PlayStation VR, 90
Pokémon GO, 4, 26, 87
Porsche, 210
Possible worlds, theory of, 244n29
Post-it notes, 109
Power grids, 192–193
PricewaterhouseCoopers (PwC), 31, 142, 199, 209
Privacy and security
   of data, 48, 56, 75, 139, 147, 151–152, 205–206, 215, 218
   managing risks of, 187
   in virtual environments, 21, 27, 42, 128–129, 138–139, 161, 181
Private sector
   banking/financial services, 153, 163–167
   commerce, 11, 31, 153–163
   consumer focus, 160–162
   fraud in, 158, 284n16
   marketing, 153–158
   media and entertainment, 153, 172–179
   and pragmatic use of metaverse, 160–161
   real estate, construction, and planning, 153, 162–163, 172–178
   travel and tourism, 159–160, 167–172
PropTech360, 172
PTSD, 150–151
Public sector
   education, 141–147, 210–211 (*see also* Immersive learning/training)
   governments, 136–141, 278n14
   health care, 135, 148–152, 160
   training programs, 139, 143–144
Pump-and-dump schemes, 158
PwC (PricewaterhouseCoopers), 31, 142, 199
"Pygmalion's Spectacles" (Weinbaum), 46

Queer Resources Directory, 241n3

Ray-Ban, 16–17, 131
Reading technologies, 80
Ready Player One (Cline), 13
Ready Player One (film), 13
Readyverse Studios, 179
Real estate, 178. *See also* Virtual real estate
Real estate, construction, and planning, 153, 162–163, 172–178. *See also* Virtual real estate

Index

Realities, 26–27. *See also* AR; MR; VR; XR
Recreational platforms' investment in metaverse, 37
Red Cross, 135
Red Hat, 119–120
Remote learning, 142–145
Remote medicine, 149–152
Remote work, 11, 32, 143–144, 184, 191, 196–198, 205–209, 212, 214, 216
Rendering, 35
Re-personalization, 5, 37, 42, 166–167
Republic of Korea (South Korea), 136–137, 164, 179, 183
Resilience, 1, 85, 89–90, 96
Reskilling Revolution (Education 4.0 Alliance), 141–142
Resolution 105, 18
REX, 172
Roberts, Ed, 79
Roblox, 13, 33, 120, 157
Robots, 196
Rodgers, Carolyn, 241n3
Rolls-Royce, 181–182
Rowling, J. K., 89
Rug pulls, 158, 284n16

Samsung, 119, 124, 182–183
Sandbox, The, 47, 136, 177, 249n22, 252n46
Santa Monica (California), 136
Scams, 158
School of Pharmacy, University of Pittsburgh, 281n44
Scott, Travis, 168, 178–179
SDGs (Sustainable Development Goals), 67, 75
Second Life, 4, 11, 13, 21–22, 182, 250n27
Security. *See* Privacy and security
Sega Genesis, 90
Sega VR, 90
Segway, 89

Self-checkout kiosks, 80
Sensorama booths, 46
Seoul, 136–137, 169
Sephora, 159
SHRM (Society for Human Resource Management), 68
Siemens, 173, 182, 201, 212
SIGGRAPH Conference, 157
Siri, 35
SK Telecom (South Korea), 179, 183, 185–186
Smart contracts, 25–26, 178
Smartphone apps, 54, 200
Smartwatches, 54, 79–80
Smith, Barbara, 241n3
Snapchat, 37
SNES, 90
Snow Crash (Stephenson), 13
Social media. *See also specific platforms*
  AI's role in, 4
  body image on, 56
  healing from, 54–56
  marginalized groups' use of, 35
  misinformation on, 57–58
  psychological impact of, 32, 54–55
  use of the term, 23
Society for Human Resource Management (SHRM), 68
Solitaire, 51
Somnium Space, 126
Sony, 119–120, 122–124, 183
South Carolina Department of Revenue data breach, 139
South Korea. *See* Republic of Korea
Spaarne (Netherlands), 149
SpaceX, 93
Spatial, 145
Spatial platform, 209
Stanford University, 142
Star Trek (TV show), 12, 52, 242n4
State of Texas data breach, 139
Statista, 126–127
Stephenson, Neal: *Snow Crash*, 13

Stereotyping, 53, 73, 88, 101, 105. *See also* Bias
Strategic alliances. *See* Partnerships
Sunk-cost fallacy, 71–72, 82
Superset, 108
Supreme Court, 125
Sustainable Development Goals (SDGs), 67, 75

Takeda, 98
Target, 159
Teachers, 143–144, 146
TechCrunch, 183
Technology
  addiction to, 53–54, 62, 223
  ecosystems of, 35–36
  emerging, 4
  energy efficiency of, 192–193
  pace of change of, 2
  reuse/recycling of digital devices, 194–195
  revolutions in, 223–224
  skepticism about, 61
  social/ethical issues with, 128, 146–147
  and telecoms, 181–184
  uniqueness of metaverse tech, 128–129
  workforce representation in, 68–69, 258n9
Technology Vision, 171
TECHSPO Technology Expo, 157
Telecoms, 179, 181–187
Telefónica, 185
Telemedicine, 149–152
Teleoperated driving, 202
Telstra, 183
Tencent, 183
Tesla, 93, 96, 124–126
3D city maps, 174–175
3M, 109
Tiffany & Co., 159
TikTok, 35

T-Mobile, 183
Tourism. *See* Travel and tourism
Toyota, 159
Toys "R" Us, 207
Toy Story (film), 91
TPG Telecom, 183
Transportation industry, 200–202
Travel and tourism, 47–48, 135–137, 159–160, 167–172, 188–189, 191
Trimble, 173
Trolling, 51, 99
Tron (film), 13
Trust, 7, 27, 56–59, 121–125
T-verse, 179
Twain, Mark, 170
Twitch, 163
Twitter (*now* X), 35, 131

UC Berkeley, 144
Ultima Online, 13
Unilever, 3
Unimersiv, 143
United Arab Emirates, 138
United Nations, 6–7, 67, 75, 135
  International Telecommunication Union (ITU), 7, 18
Unity, 120, 124, 196–197
Universes, theory of multiple, 19, 244n29
Universities, metaverse's impact on, 145–146
University of Maryland, 144
University of Massachusetts, Amherst, 191
University of Pittsburgh Medical Center (UPMC), 149, 281n44
University of Southern California, 145
UPS, 210
Urban planning. *See* Real estate, construction, and planning
USB technology, 119
User-focused design, 3, 5
US Patent and Trademark Office, 83

Vans, 157
Verge, The, 183
Verizon, 183, 185
VictoryXR, 145
Videoconferencing, 36–37
Video games, 4. See also Avatars;
  Gaming; Second Life
Video telephony, 12
Virtual Boy, 85, 90
Virtual classrooms, 142–145
Virtual design and construction (VDC), 173
Virtual embassies, 138
Virtual factories, 107
Virtual People course, Stanford University, 142
Virtual real estate, 17, 41–43, 162–163, 173–174, 177, 217. See also Real estate, construction, and planning
Virtual reality. See VR
Vision Buddy, 105
Vision Pro, 127
VMware, 120
Vodafone, 179, 183, 185
Voice assistants, 35, 80
Volkswagen, 196
Voter Database data breach, 139
Voxels, 177
VR (virtual reality), 4
  birth of, 46
  consumer interest in headsets, 126
  defined, 26
  failures and risk-taking in, 84–87
  in fiction and science, 13, 15, 46
  gaming's role in headset development, 14
  headsets for, 14, 16–17, 26, 46, 84–85, 127 (see also Oculus Quest; Oculus Rift)
  health care use of, 97
  investment in, 155
  vs. metaverse, 14
  notable failures in, 90

popularity of, 11
training in, 165
and users with mobility challenges, 67
VRChat, 126

Walled gardens, 27
Walmart, 14, 159, 201
Warby Parker, 159
Warehouse management and logistics, 200–202
Warner Bros., 124, 179
Web 1.0, 45–46
Web 2.0, 19–20, 45, 54–56, 58, 161–162, 176. See also Social media
Web 3.0. See also Metaverse
  emergence of, 15
  vs. the internet, 52–53
  and metanomics, 43
  network improvements, 35
  sensory and immersive experiences enhanced by, 33
  virtual worlds supported by, 27
WEF (World Economic Forum), 67, 141
Weinbaum, Stanley: "Pygmalion's Spectacles," 46
Wells Fargo, 28, 163–164
Wheatstone, Charles, 46
Whose Metaverse?, 142
Wi-Fi, 119
Windows 3.0, 51
Windows laptops and tablets, 119
Windows Phone operating system, 89
Work, future of, 205–222. See also Remote work
  Bank of America (case study), 208, 211–212
  communication and collaboration, 207–208
  cost savings and managerial tools, 216
  customer service and training, 209–212

Work, future of (cont.)
  cybersecurity, 216
  digital economy, 216–217
  digital well-being and mental health, 213–214
  employee autonomy and well-being, 212–213
  ethics and governance, 215, 219, 221–222 (*see also* Privacy and security: of data)
  generative AI, 217–218
  landscape of, 205–206
  leaders, role of, 214–216
  management and accountability, 207–208
  opportunities, 220–221
  overview of, 219–220
  professional roles, 219
  risks, 220–221
  and scalability, 218
  skill building and lifelong learning, 210–211, 219
  virtual teams and organizational culture, 214–215
  work–life balance, 206, 212, 214
Workforce growth, 68–69
World Economic Forum (WEF), 67, 141
World Health Organization, 135
World hopping, 185
World Intellectual Property Organization, 83
World of Warcraft, 13–14, 247n7
World Telecommunication Standardization Assembly (WTSA-24), 18
World Wide Web, invention of, 45, 241n1. *See also* Web 1.0; Web 2.0; Web 3.0
World Wildlife Fund, 135

X (*formerly* Twitter), 35, 131
Xbox Live, 124
Xerox, 119

XR (extended reality), 24, 26–27
XRHealth, 150
XR World, 179, 185

Yahoo, 47
Your VR Therapy, 150

ZELF, 165–166